IMAGES
of America

NORWICH FIREFIGHTING

On the cover: In 1922, the Norwich Fire Department converted all firefighting apparatus to automobiles. This giant leap in technology marked the end of the horse-drawn era and the beginning of modern firefighting. The photograph was taken in 1927 outside fire headquarters, located at 24 Chestnut Street. (Courtesy of the Norwich Fire Department.)

IMAGES
of America

NORWICH FIREFIGHTING

Glenn M. Watts

ARCADIA
PUBLISHING

ISBN 978-0-7385-5549-2

Published by Arcadia Publishing
Charleston SC, Chicago IL, Portsmouth NH, San Francisco CA

Printed in the United States of America

Library of Congress Catalog Card Number: 2007933903

For all general information contact Arcadia Publishing at:
Telephone 843-853-2070
Fax 843-853-0044
E-mail sales@arcadiapublishing.com
For customer service and orders:
Toll-Free 1-888-313-2665

Visit us on the Internet at www.arcadiapublishing.com

This book is dedicated to the nine individuals of the Norwich Fire Department who lost their lives while serving the city of Norwich. These men will always be remembered and honored.

Pvt. Adam Ensling (February 11, 1942)
Pvt. Daniel Falvey (May 5, 1945)
Chief Joseph Patrick Tracy (September 9, 1951)
Pvt. Frank William Wasik (February 15, 1952)
Pvt. Leonard Malcolm Counihan (April 3, 1962)
Pvt. Edward Romano (April 3, 1962)
Pvt. Carl Joseph Burke (April 3, 1962)
Capt. William John Sheridan (April 3, 1962)
Chief Charles Francis Matthews (November 16, 1967)

Contents

ACKNOWLEDGMENTS

I would like to thank the following individuals and groups for their support and contributions: Stella Moran; Faith Jennings; the following from the Norwich Fire Department Chief Howard L. Stanton, Chief Harold Lamphere Jr., Chief Thomas Kirby, Chief James Walsh, Chief Kenneth Scandariato, Deputy Chief Calvin Cobb, Deputy Chief Ernest Champagne, Deputy Chief Kenneth Merkett, Deputy Chief John Donahue, Deputy Chief William Tessier, Deputy Chief Thomas Casey, Capt. Raymond Selvidio, Capt. John Mathieu, Capt. Stanley Homski Jr., Lt. Stuart Carter, Lt. Joseph Hebert, Lt. Scott Merchant, Lt. Keith Wucik, Pvt. Peter Davis, Pvt. Ronald LePage, Pvt. Raymond S. Benoit, Pvt. Paul Tetreault, Pvt. Michael Tirone, Pvt. Stephen Fricke, Pvt. Michael Dolan, Pvt. Scott Suplita, and Pvt. Gail Noyes; Bill Stanley; Dale Plummer; Gary Pinkham (Pfizer Fire Department); Bill Eyberse (Yantic Fire Engine Company No. 1); William McInerney (Middletown Fire Department); Leonard Luzusky (Manchester Eighth District Fire Department); Leon Barnowski (Norwich Management Information Systems Department); Connecticut Fireman's Historical Museum; and Norwich Firefighters Local 892. *Norwich Firefighting* could not have been completed if it were not for their kindness and generosity.

INTRODUCTION

The Norwich Fire Department has a rich and historic tradition of service to the city of Norwich. Today's career fire department boasts an impressive 105 years of service with its formative volunteer history dating back almost 200 years to the beginning of the 1800s.

Norwich Fire Department evolved from 11 separate volunteer fire companies in the Central Consolidated District, also known as the city of Norwich. This three-and-a-half-square-mile sector encompasses the sprawling landscape around Norwich Harbor and was the major epicenter of the town during the early 1800s. Growth of the city and the influx in population were largely due to the manufacturing and textile mills that sprung up along the many waterways of the downtown area during the Civil War. Industrial Revolution pioneers exploited the Yantic, Shetucket, and Thames Rivers to manufacture textiles and munitions for the Union soldiers of the war. With increased jobs came larger populations, additional housing, and larger buildings used for storage and commerce.

In 1876, the rapid growth of the city necessitated a permanent fire chief and fire marshal to oversee fire protection and organize the many autonomous fire companies of the downtown area. Chief Joseph B. Carrier was appointed permanent fire chief of the Greeneville and Norwich Fire Departments in 1876 and served as fire marshal and alarm superintendent until his death in 1890. Carrier's appointment to this career position paved the way for the permanent force that started in 1902 under the leadership of then chief Howard L. Stanton.

The present-day Norwich Fire Department is proud to honor its past while continuing to uphold its tradition of service to the inhabitants of the Central Consolidated District.

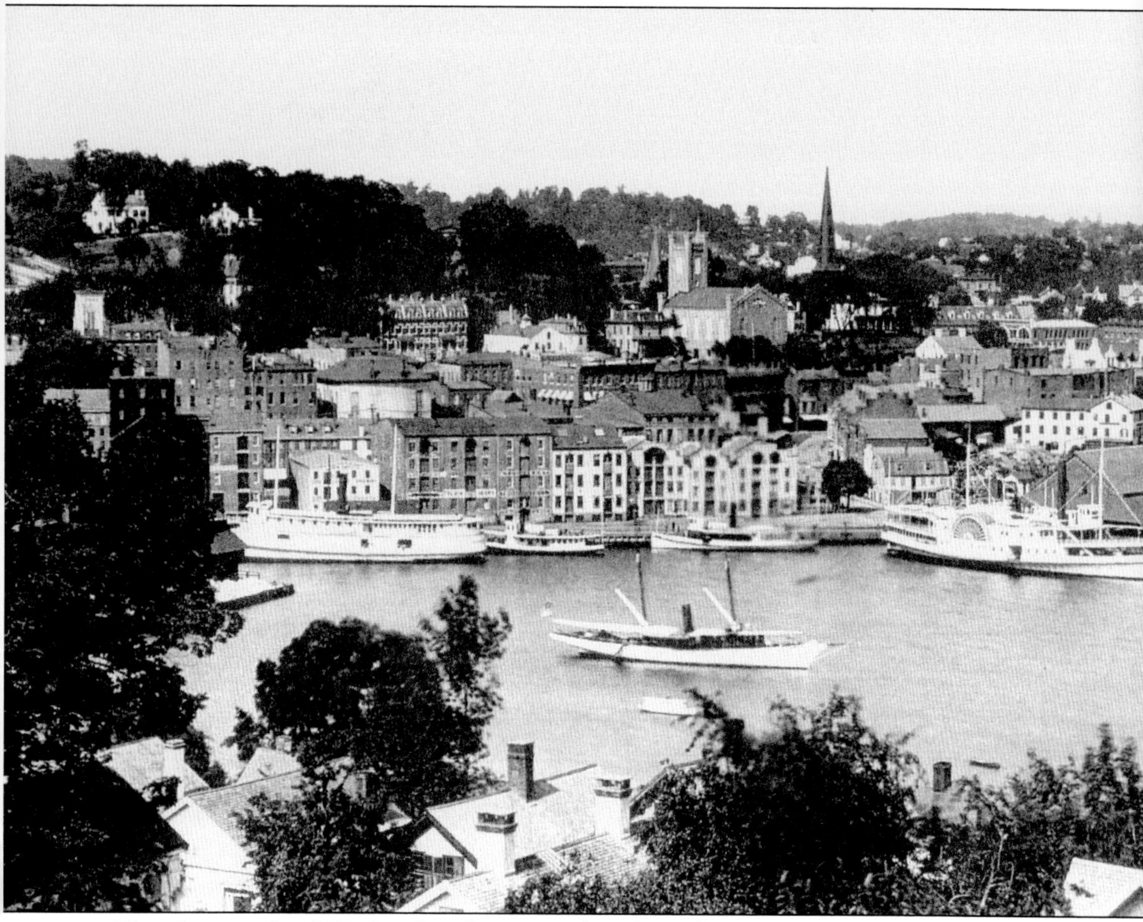

This photograph of Chelsea Landing was taken from the city's west side around 1900. Many homes that occupied the west bank of the Thames River, observed in the foreground, were demolished during the 1970s as part of the west side revitalization.

One

The Volunteer Era

Norwich Fire Department's roots began around 1804, with the organization known as the "Red Jackets." This group of 25 to 40 influential residents of Chelsea Landing, the present-day center of the city, gave rise to a volunteer force that would eventually include 11 separate fire companies.

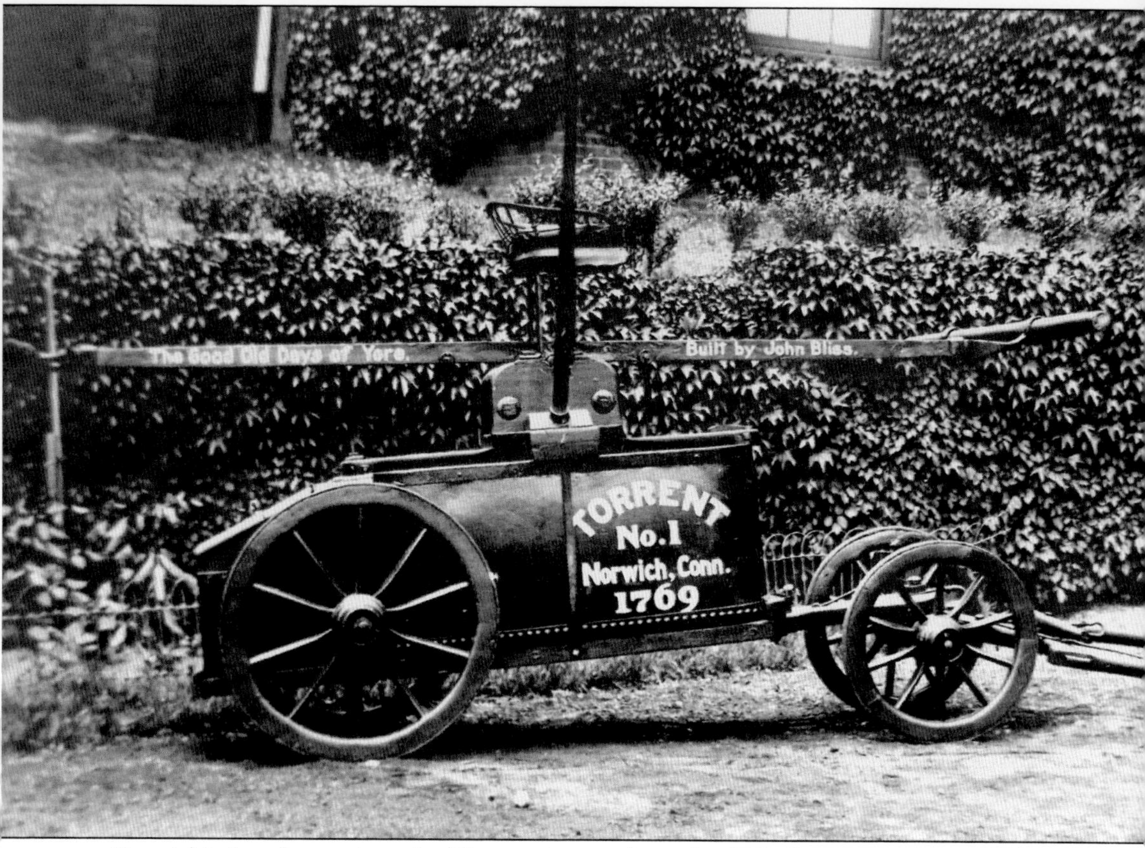

Norwich's first fire engine, the *Torrent*, is seen here. In 1769, Stonington native John Bliss constructed a handmade fire apparatus out of chestnut, butternut, oak, and pine. This beautifully crafted fire engine was placed in service in Norwichtown at the local blacksmith's shop. The wooden wheels were treaded with steel bearing the taxation seal of England's king. The 110-gallon tub was hand riveted and made of hand-peened copper. Four to six men were needed to pump the apparatus while an additional 12 or more were needed to keep the tub filled using water buckets. Leather clapper valves weighted with lead made it possible for a group of men to deliver a pressurized stream of water 150 to 175 feet. The Torrent was in service for roughly 100 years in Norwichtown. Following its many years of service, the *Torrent* visited many venues across the country, including the 1904 world's fair in St. Louis, Missouri. It is regarded as the sixth-oldest American-made horse-drawn hand engine in existence today and presently resides at fire headquarters, located at 10 North Thames Street in Norwich.

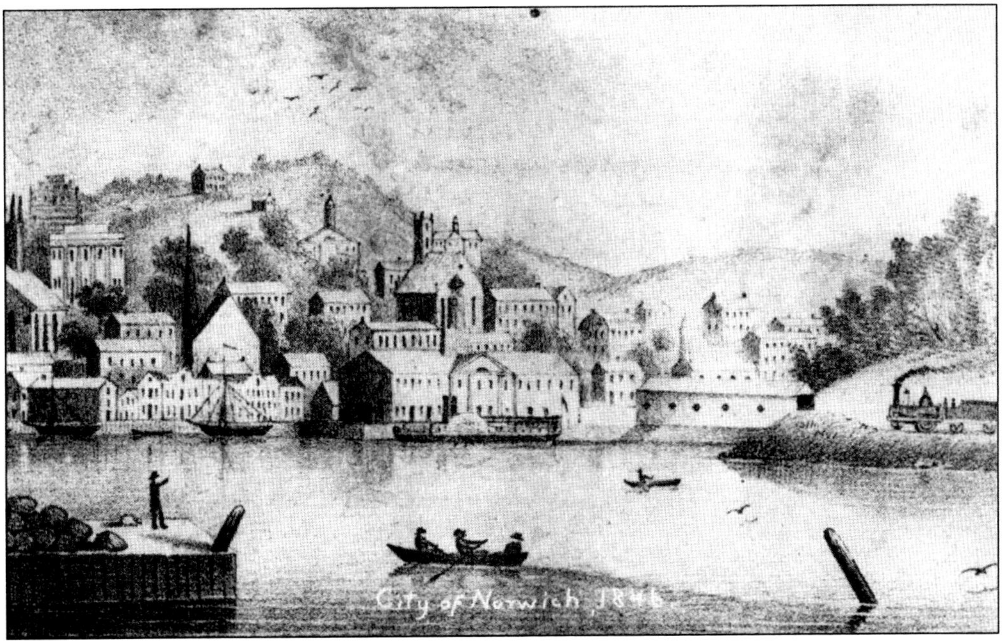

In 1846, the banks of the Thames River began to blossom with dwellings and industry. As Chelsea Landing grew, so too did the need for fire protection. On August 15, 1845, Col. James Spaulding was appointed chief engineer of the Norwich Fire Department by the court of common council. In addition to Spaulding's appointment, two assistant engineers and 12 fire wardens were named to the department.

At a meeting of the Court of Common Council of the City of Norwich, held August 15th, 1845, the following appointments were made.

ENGINEERS OF FIRE DEPARTMENT.

Col. James Spaulding, Chief Engineer.
Alba F. Smith, 1st Assistant Engineer.
Albert H. Benchley, 2d Assistant Engineer.

FIRE WARDENS.

Ward No. 1.—Amos W. Prentice, Joshua W. Shepherd, E. P. Miner, John B. Perry 2d.

No. 2.—John G. Huntington, Rufus L. Fanning, James B. Bromley, Edward Chappell.

No. 3.—Daniel Bassett, Amos E. Cobb, Asa Manning, William Phillips.

NORWICH CEMETERY

Is situated in East Chelsea, on Cliff Street; established in 18—; it is divided into lots of convenient size, and was all sold so that it became necessary to establish a new one, which was done by the establishment of the

YANTIC CEMETERY,

On Yantic Street above the Falls, in 1844. The ground is beautifully situated on Yantic River, and tastefully laid out and ornamented. It is divided into lots of convenient size, many of which are purchased and some are occupied. The Lot for the purpose for which it is designed is a delightful place.

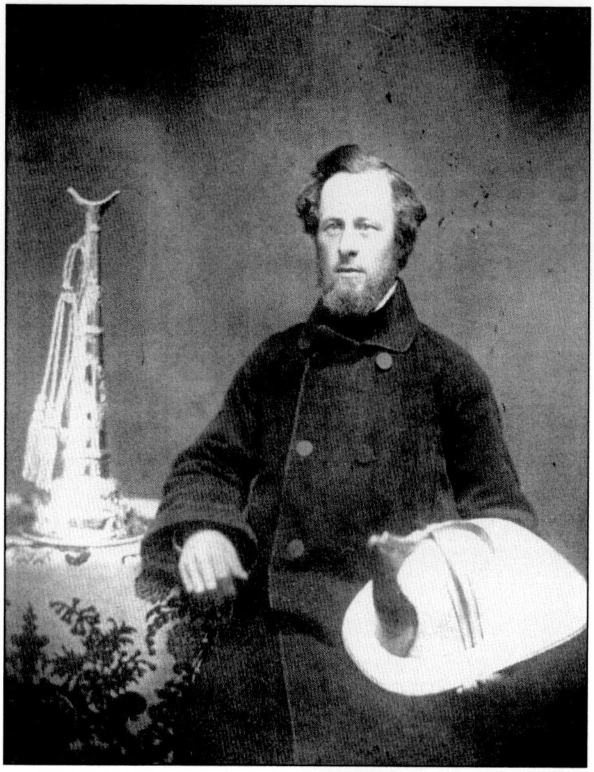

Edward S. Osgood was appointed chief engineer of the department in 1857. The following year, he resigned from the position and continued to serve the department as the foreman for the Wauregan Steam Fire Engine Company No. 1 until 1862.

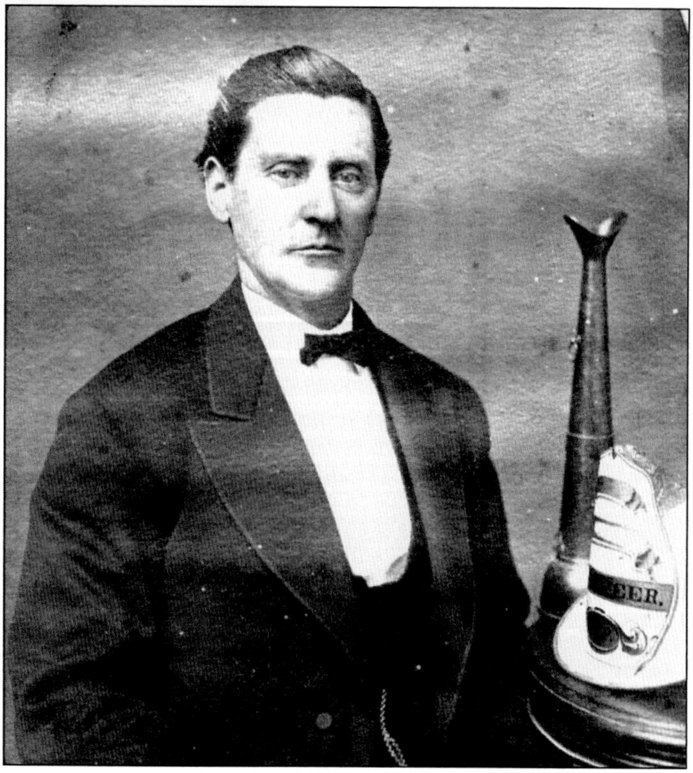

Chief Engineer Henry Peale is shown here around 1860. Peale served as chief engineer for a period of two years before being succeeded by Sylvanus Downer in 1862.

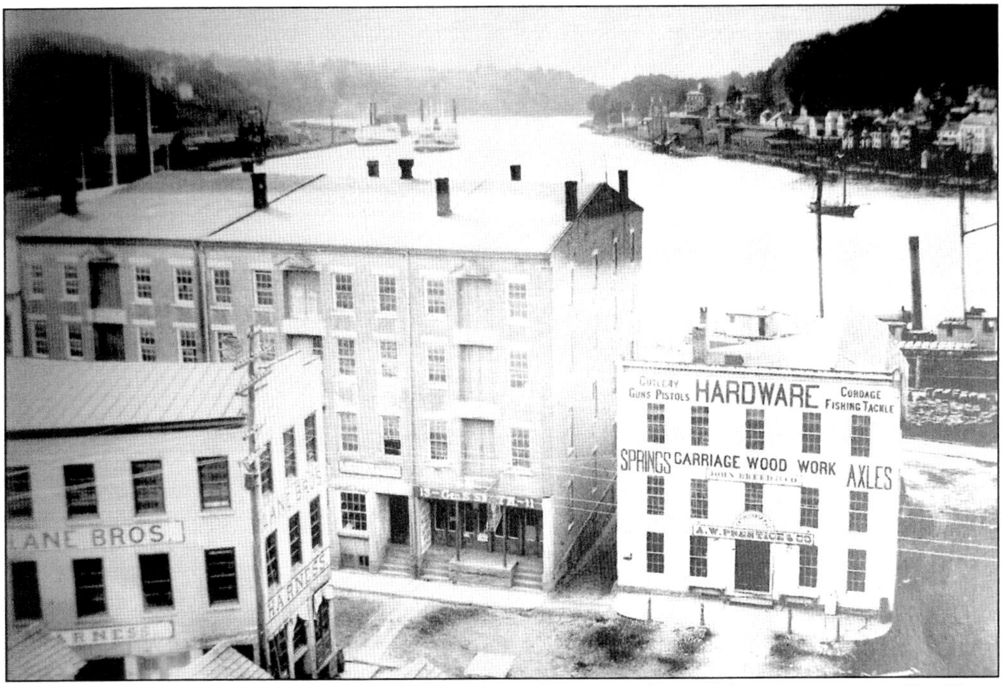

This photograph of Chelsea Landing, taken during the 1800s, depicts the commerce and building structures of the downtown harbor area. Fire hazards in these heavy-timber buildings spurred the organization of additional fire companies throughout the mid-1800s.

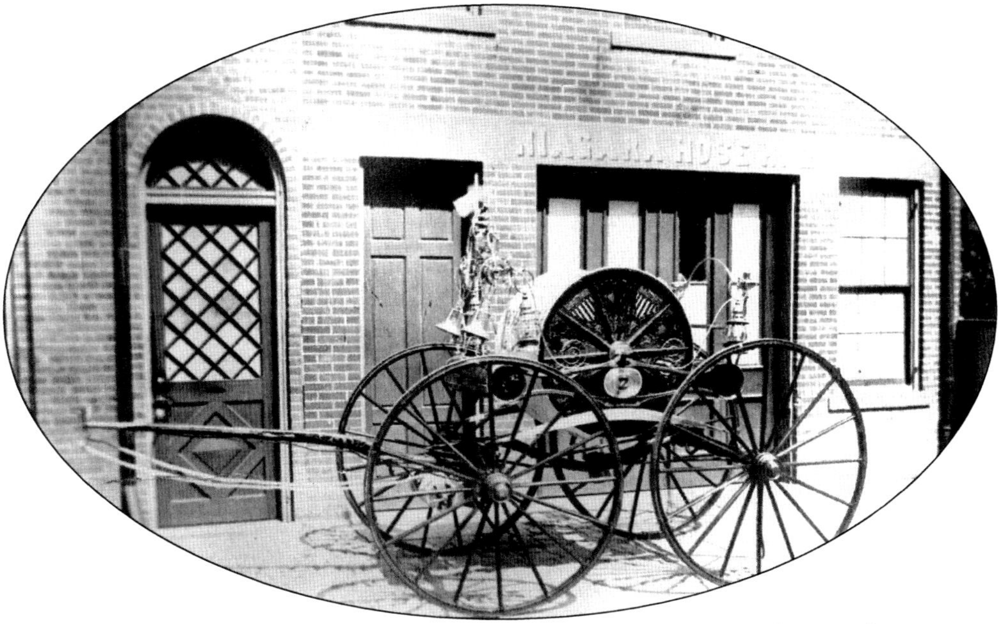

Originally organized in 1815 on Church Street, Niagara No. 2 eventually moved quarters to lower Union Street in 1849. The lower portion of Union Street would later become part of Broadway. In 1872, Niagara No. 2, originally a hand engine company, was reorganized as a hose company and existed until the organization disbanded in 1903 to make room for the newly formed paid department.

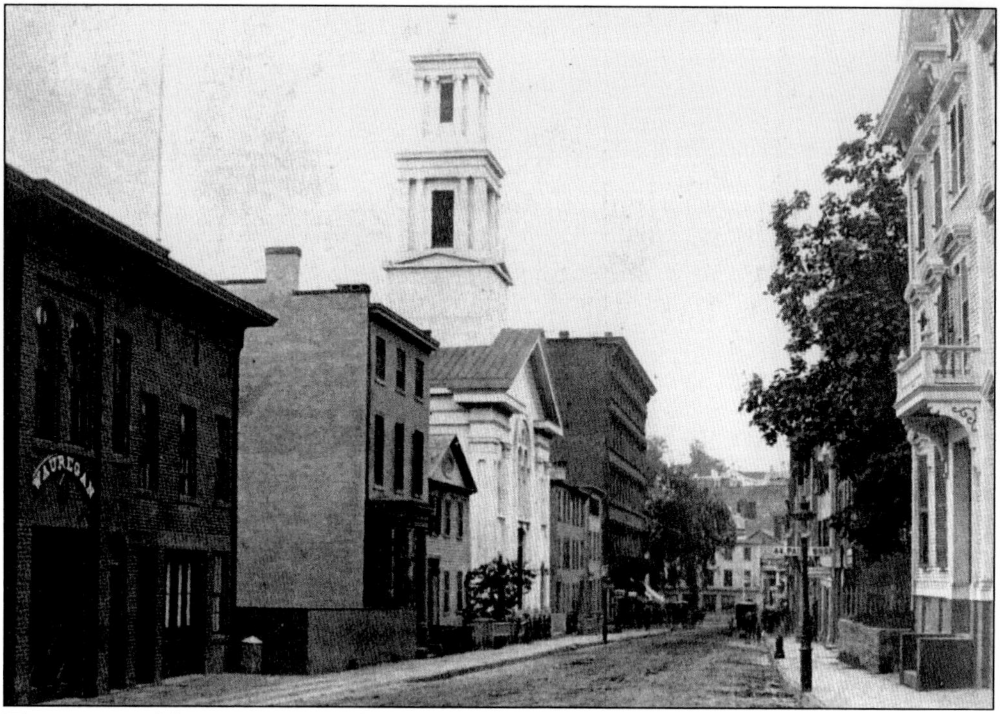

The quarters of the Wauregan Hook and Ladder Company No. 1 and the Niagara Hose Company No. 2 can be seen in this photograph of lower Broadway around 1880.

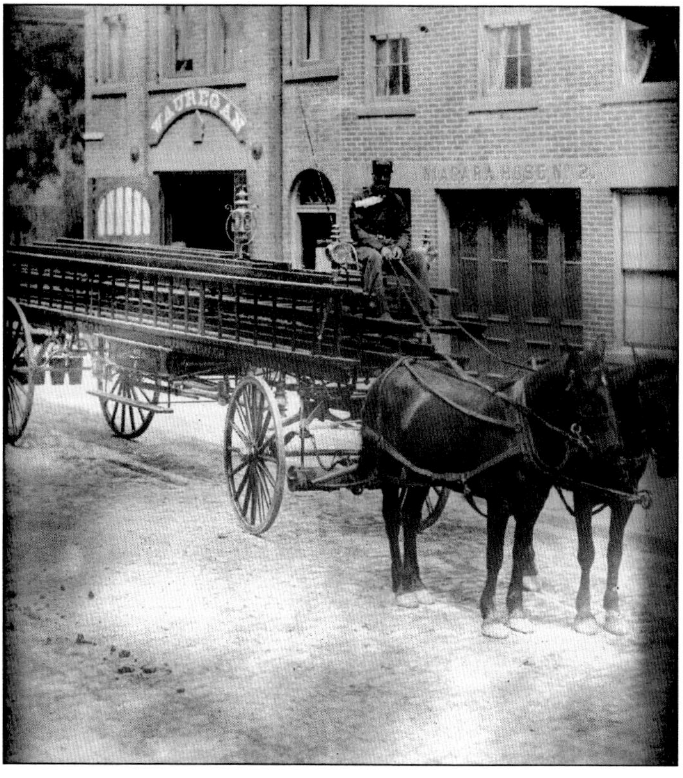

Originally organized as the Waverly Hook and Ladder Company around 1858, the Wauregan Hook and Ladder Company No. 1 was formed in 1861 and was located at 31 Union Street adjacent to Niagara Hose Company No. 2. The address of this firehouse would later become 67 Broadway. The Wauregan Hook and Ladder Company No. 1 remained in service from 1861 until 1905 when the company disbanded.

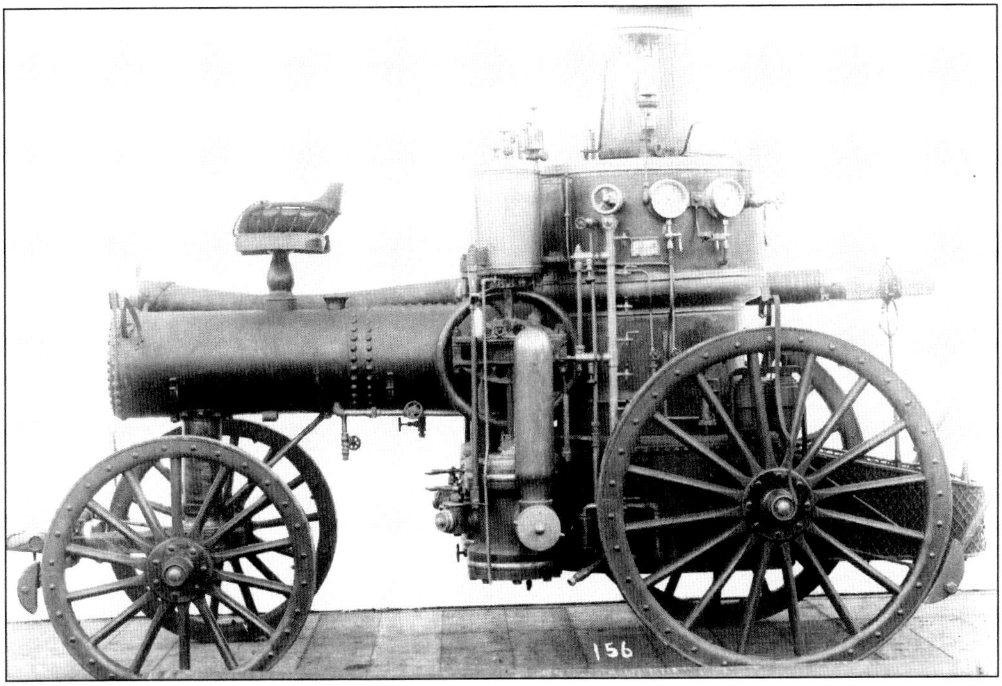

The *Wauregan* (war-e-gan) was the first steam engine purchased by the Norwich Fire Department. The engine was constructed in 1861 by the Amoskeag Locomotive Company of Manchester, New Hampshire. The apparatus weighed 9,600 pounds and had a capacity of approximately 700 gallons per minute. This monster steam engine was located at 67–69 Main Street and was in existence from 1861 to 1905 when it was traded as partial payment for the career department's new American LaFrance Metropolitan steamer. The name *Wauregan* is derived from the language of the Mohegan Indians, indigenous to the Norwich area. The name is loosely translated "everything well and good."

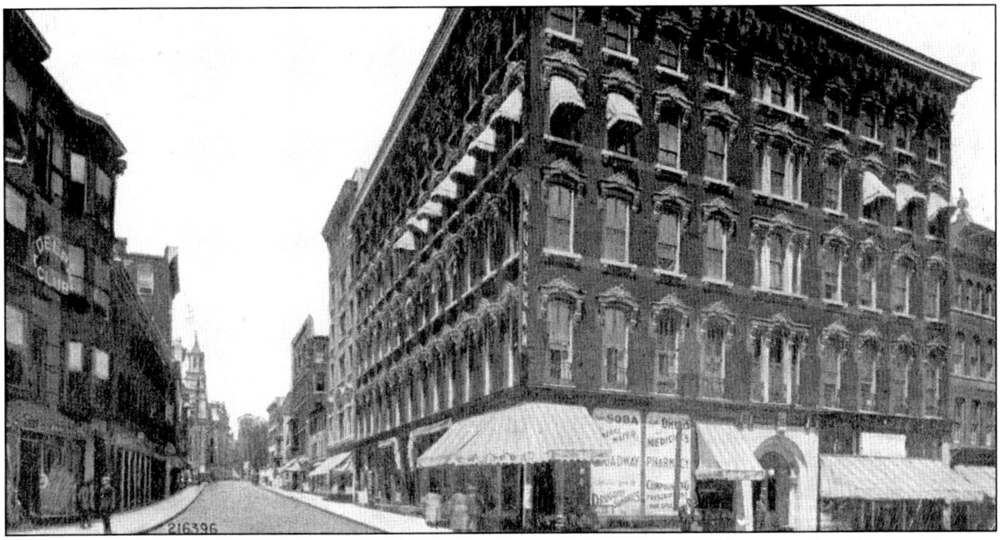

The Wauregan Hotel was erected in the late 1850s and was revered by travelers for its elegant and upscale accommodations. The hotel was located at the corner of Main Street and Broadway and was the inspiration for the two Wauregan fire companies that were formed in 1861.

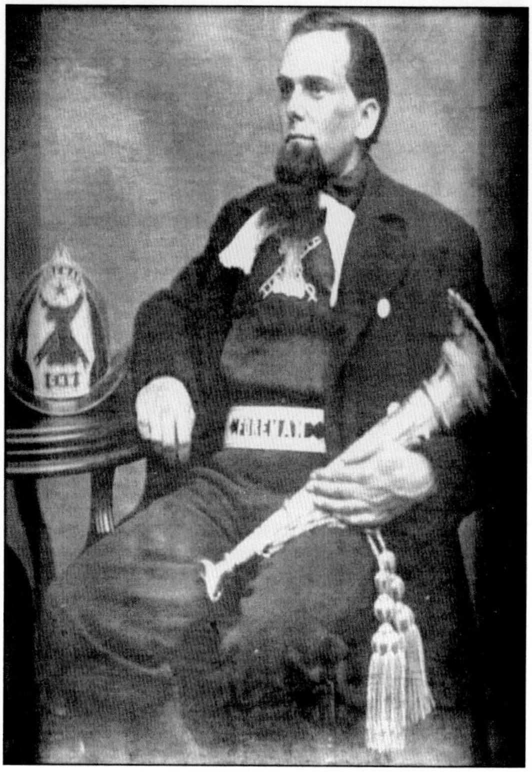

Charles H. Whaley served as the foreman for the Wauregan Hook and Ladder Company No. 1 from 1867 until 1870.

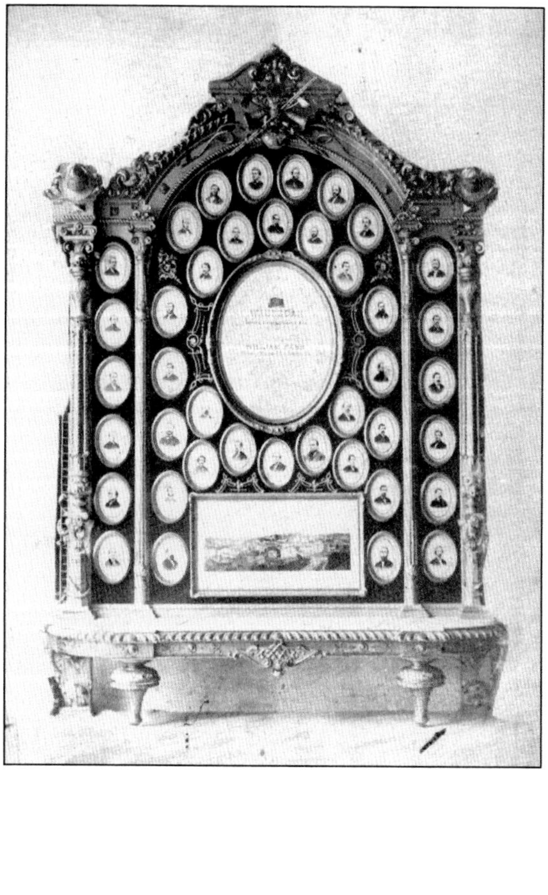

This photograph of a framed picture captures the grand scale of presentation awards and gestures of volunteer fire companies in the 1800s. The Wauregan Hook and Ladder Company No. 1 presented this 8.5-foot-by-11-foot gilt frame to the William Penn Hose and Steam Fire Company of Philadelphia, Pennsylvania. This is the second known frame of this size, with the first frame on loan to the Connecticut Firemen's Historical Museum in Manchester.

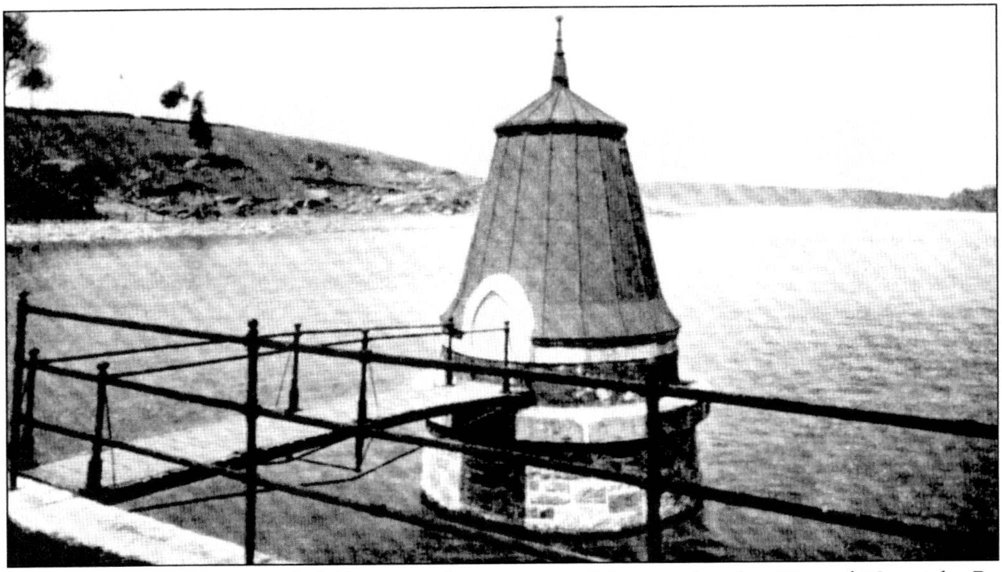

The waterworks project of 1868 was a large accomplishment by the City of Norwich. By impounding tributaries in the hills, large reservoirs were created. Due to the reservoirs' elevation above sea level (250 feet in 1885), these water bodies were able to generate pressurized water to the newly installed city hydrants. In later years, pressures in upwards of 85 pounds per square inch were generated downtown. This monumental achievement for fire protection was organized by then chief William M. Williams. The pressurized hydrant system of the downtown area would eventually give rise to additional hose carts in the fire department.

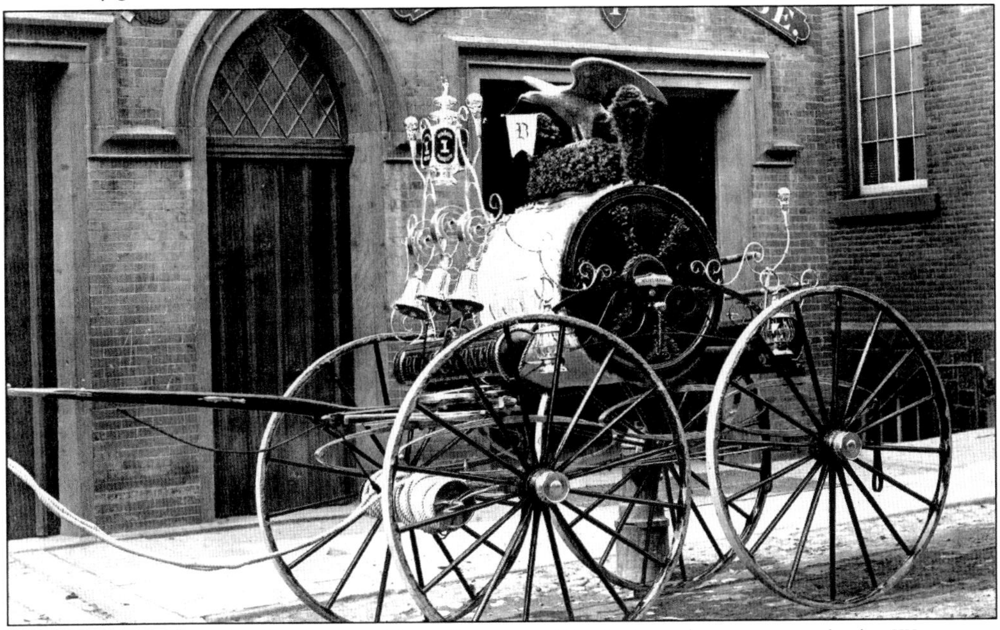

Organized in 1868, the Blackstone Hose Company No. 1 shared quarters with the Wauregan Steam Fire Engine Company No. 1 on lower Main Street. The craftsmanship and detail of the 1800s fire apparatus is personified in the picture above. Polished metal, decorative glass, and beautiful etchings adorned the apparatus of the era. Blackstone Hose Company No. 1 was in service from 1868 to 1903 when the company disbanded.

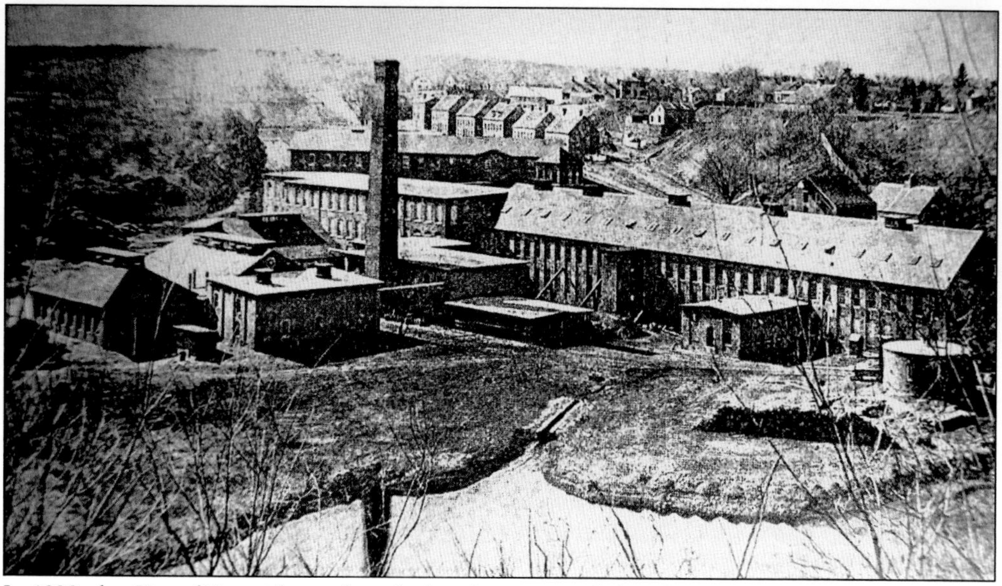

In 1828, the City of Norwich purchased a hand engine and placed it at the Falls Cotton Company located on Yantic Street. This fire company was known as No. 3 and later as Uncas Engine Company No. 3 after the Mohegan Indian Sachem who had ties to the falls that powered the mill.

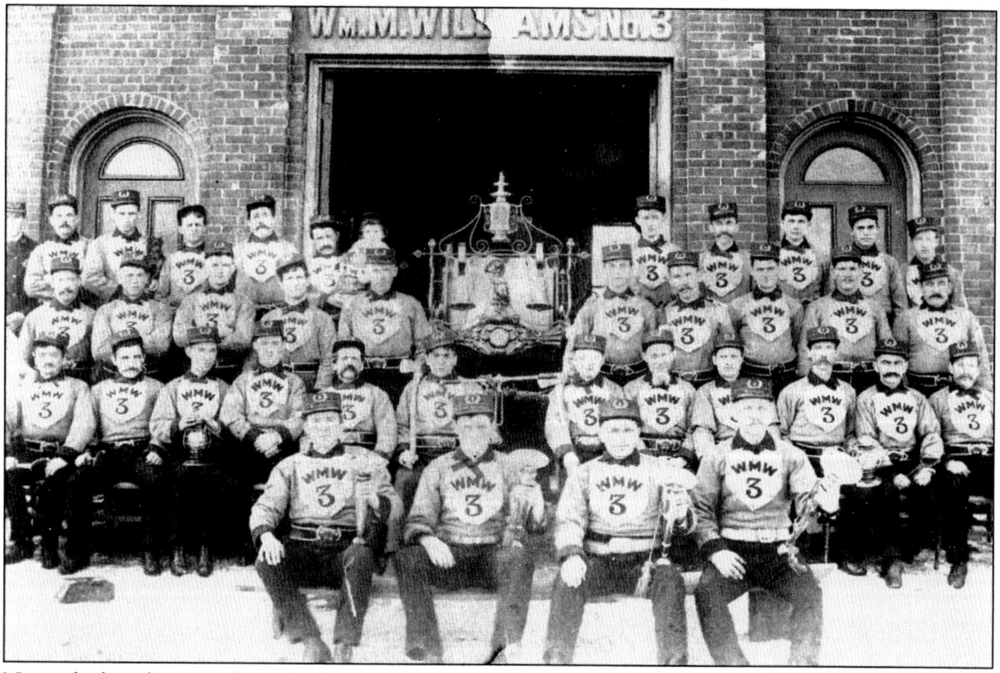

Named after the presiding chief engineer, William M. Williams, this newly formed fire company of the same name was organized in 1867 and located at 45 Sherman Street. Originally organized as a steamer company, William M. Williams No. 3 operated at that capacity until 1902 when the steam engine was replaced with a hose cart. This steamer was one of three new machines purchased by the city in 1867. Built by William Jeffers of Pawtucket, Rhode Island, the single-acting pump was capable of 300 gallons per minute. The photograph above shows the men of the William M. Williams Hose Company No. 3 in full dress uniform around 1902.

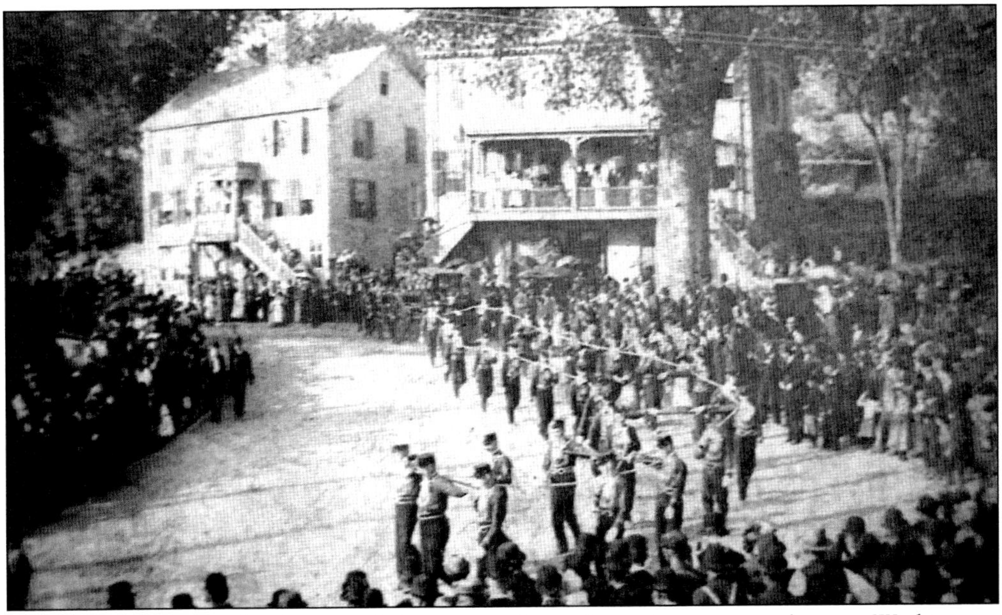

Grand parades typically showcased fire companies such as the one seen here at Washington Square. Many of the organizations of this era were socially linked and often operated independently. Competitions called musters were not uncommon at parades and festivals. Fire companies were often pitted against one another in highly competitive disciplines with nothing more than bragging rights as their reward.

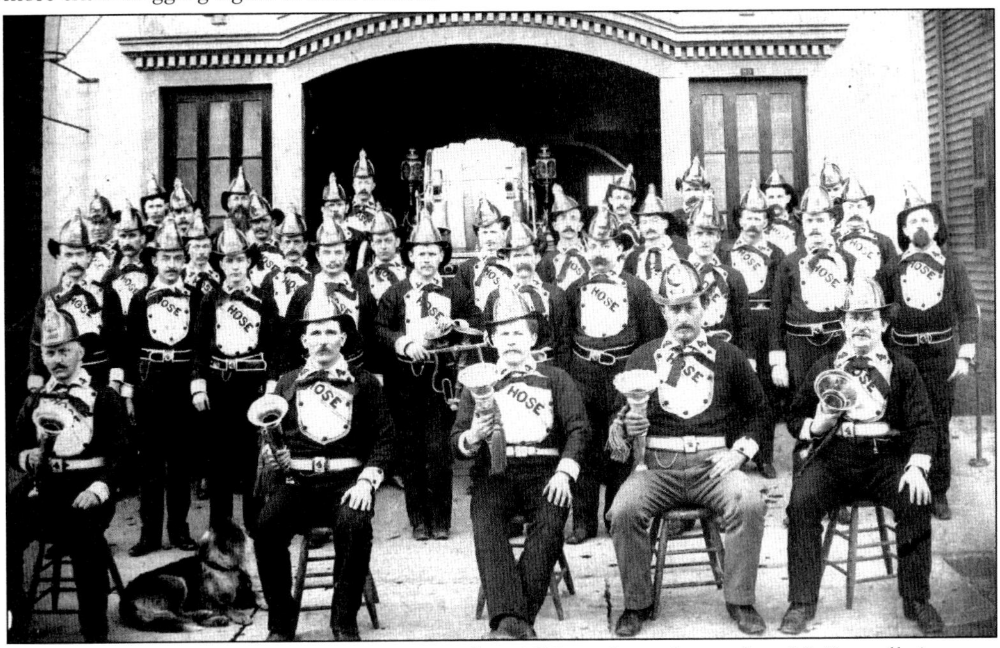

Norwich Hand Engine No. 4 was organized in 1872 and was located at 80 Boswell Avenue. Norwich Hand Engine No. 4 was in service from 1872 until 1874 when the organization was changed to a hose company. Norwich Hose Company No. 4 provided fire protection to the city of Norwich from 1874 until the fire company disbanded in 1906. The men of Norwich Hose Company No. 4 are pictured here in full dress uniform.

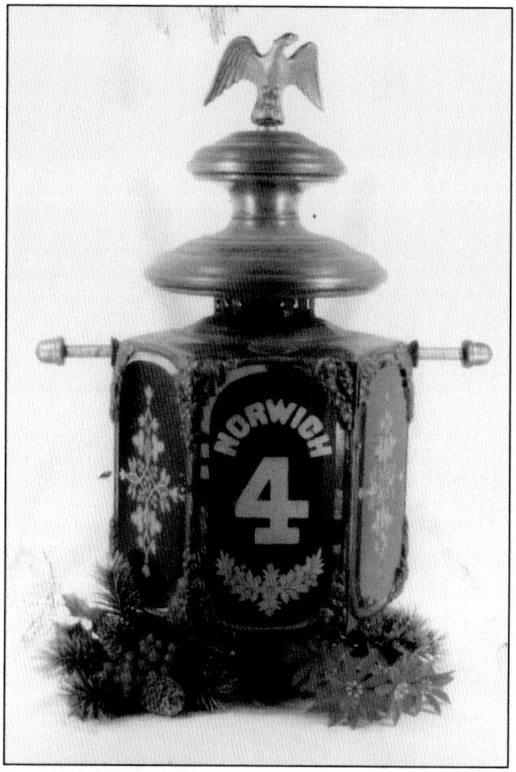

Some fire lanterns of the era were functional while others were purely decorative. This glass lantern from Norwich Hose Company No. 4 is on loan to the Connecticut Firemen's Historical Museum in Manchester.

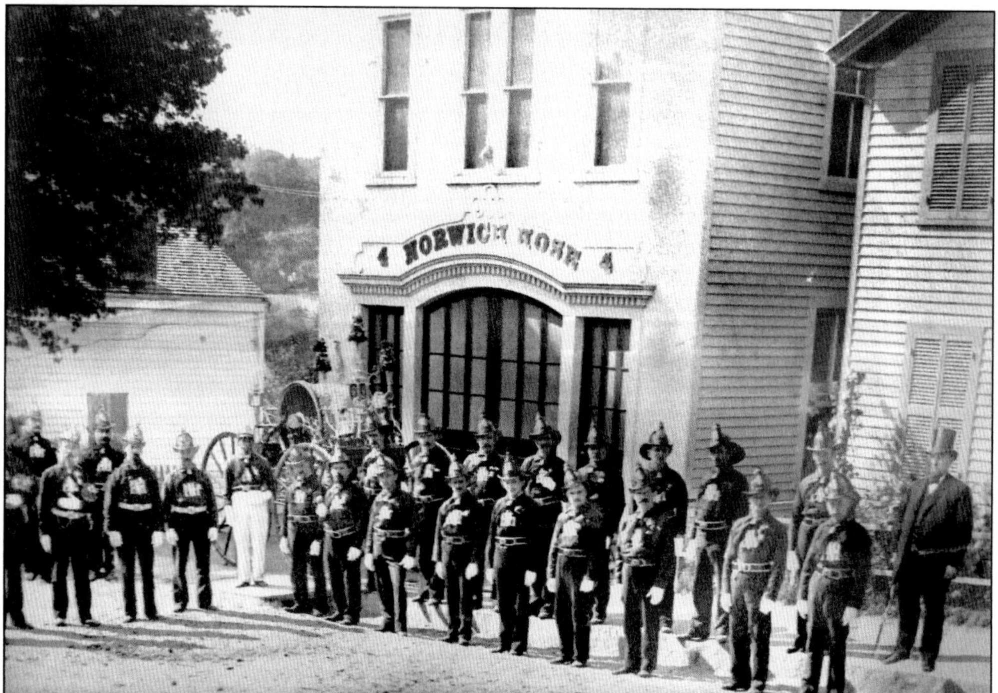

The men of Norwich Hose Company No. 4 stand with their apparatus outside their quarters at 80 Boswell Avenue.

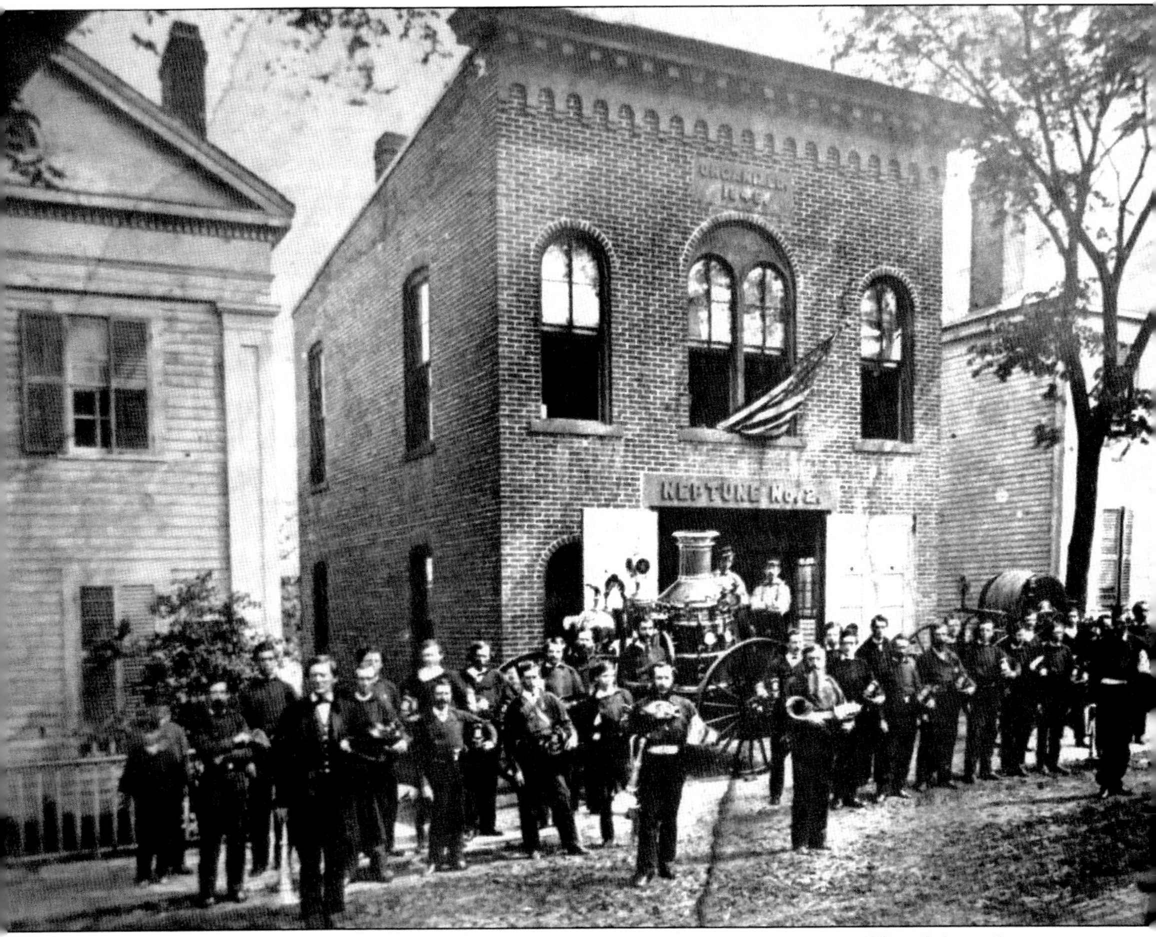

Neptune No. 2 is pictured here around 1867 outside its quarters at 181 West Main Street. Originally organized in 1846 as Neptune No. 5 on Thames Street, the company was relocated to its new home at 181 West Main Street in 1859. Neptune's designation of No. 5 was changed to No. 2 in 1867 when it received one of the city's three newly purchased steam engines. Neptune Steam Fire Engine Company No. 2 operated from this firehouse until 1902 when the organization disbanded.

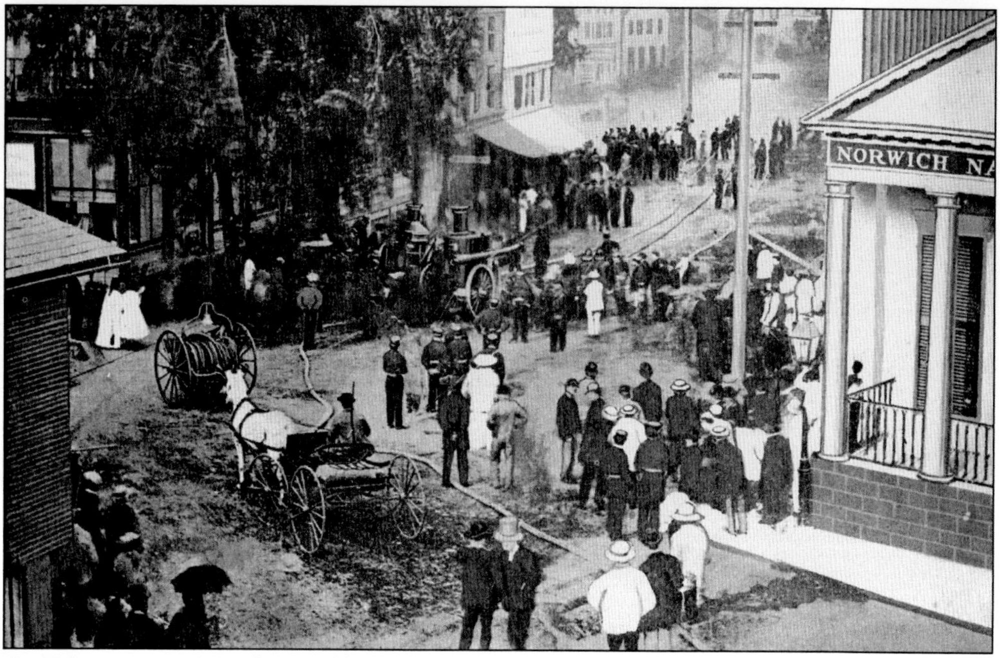

On February 14, 1869, the Apollo Hall was destroyed by a large fire that began at 12:30 a.m. in an adjacent structure. The fire was one of the largest in many decades and consumed the downtown landmark. In this photograph, a hose wagon and two steam engines are pictured operating at the fire, which was located on Main Street opposite the Wauregan Hotel.

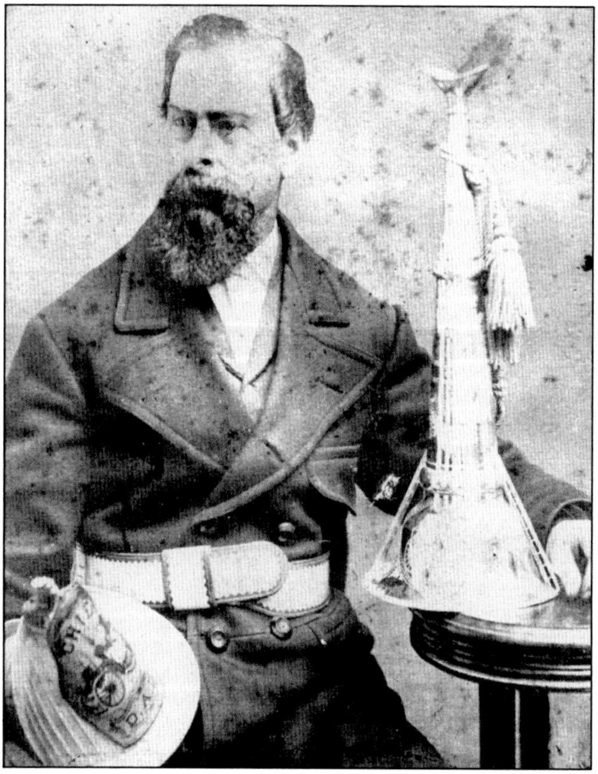

Daniel A. Delanoy was appointed chief engineer by the court of common council in 1871. He served as chief from 1871 to 1873 before being succeeded by Joseph B. Carrier in 1874. During his tenure as chief, the Neptune Steam Engine Company No. 2 was renamed Delanoy Steam Engine No. 5. The steam engine held this title until 1880 when the name was changed back to Neptune.

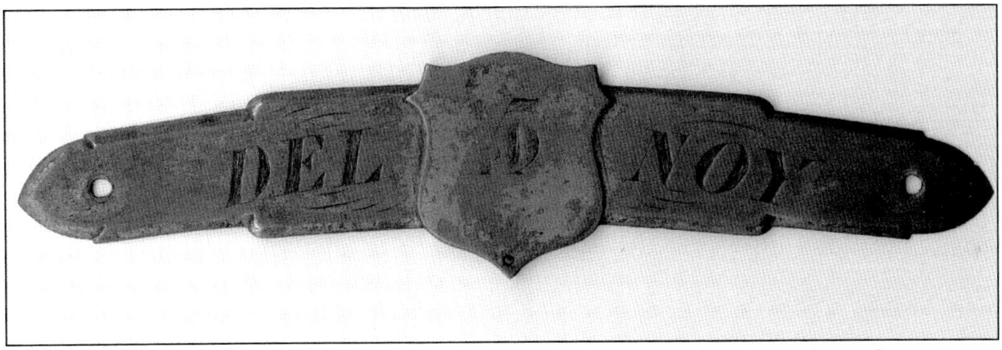

This photograph shows the boiler plate of Delanoy steam engine No. 5. The Delanoy name came from the acting chief of the department in 1871 and was used for approximately eight years before the name was retired. The photograph of the boiler plate clearly shows the etched letters of Delanoy and the intertwined No. 5 on the shield.

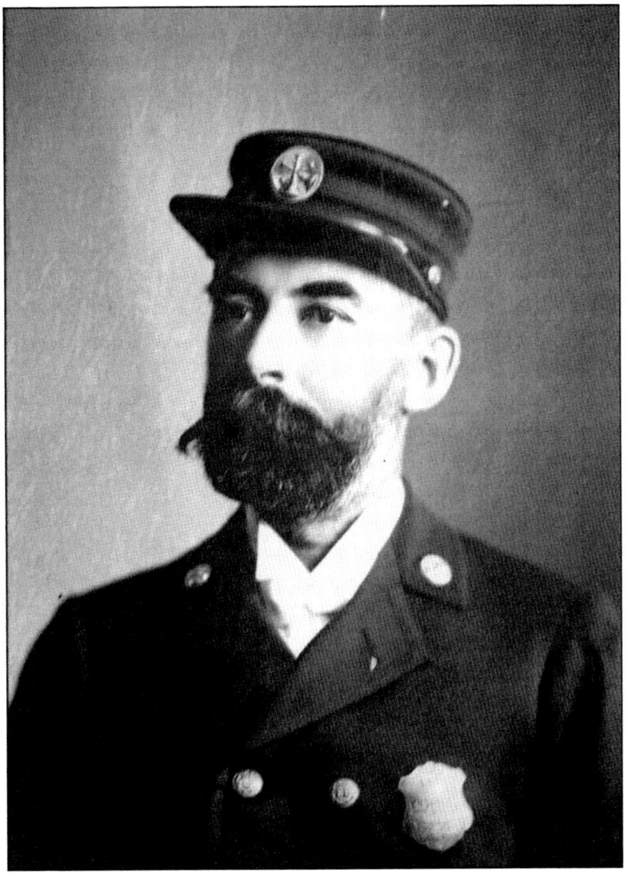

First Assistant Chief Howard L. Stanton is pictured here around 1890. Prior to his appointment to assistant chief, Howard L. Stanton was foreman of the Neptune Steam Engine Company.

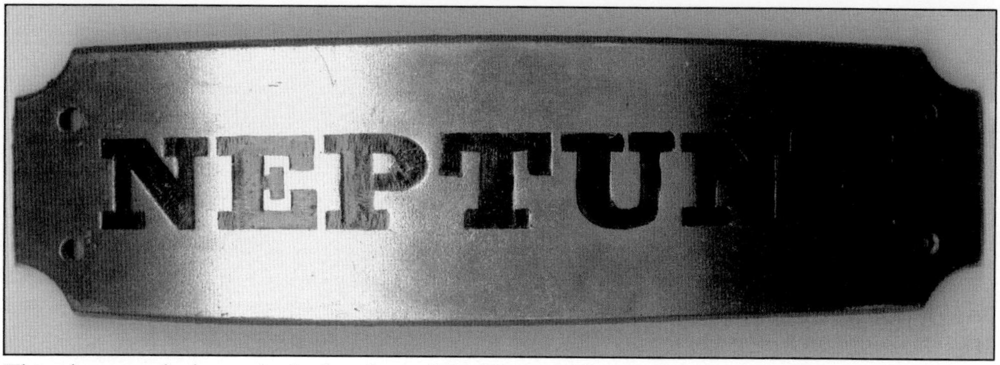

This photograph shows the boiler plate of the Neptune Steam Engine Company.

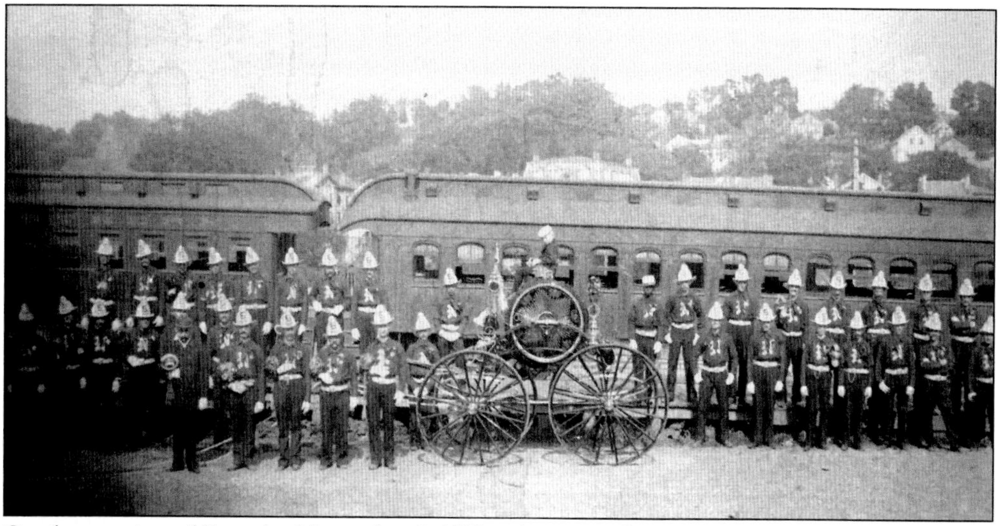

On the evening of Saturday November 9, 1872, a large conflagration in Boston, Massachusetts, began to spread so rapidly that nearby communities were called for aid. The Honorable Mayor James Lloyd Greene of Norwich asked for assistance from the Norwich Fire Department. Approximately 100 men along with dignitaries and prominent individuals from Norwich boarded a train on the Norwich/Worcester Rail Road. It was decided that the Wauregan steam engine, Delanoy steam engine and hose carriage from the Wauregan Hook and Ladder Company would be shipped with the men. The Norwich firemen helped battle the large conflagration over the next few days and returned the evening of the 11th. The photograph above, taken a few years after the great Boston Fire, shows the men of Neptune Hose Company at the train station.

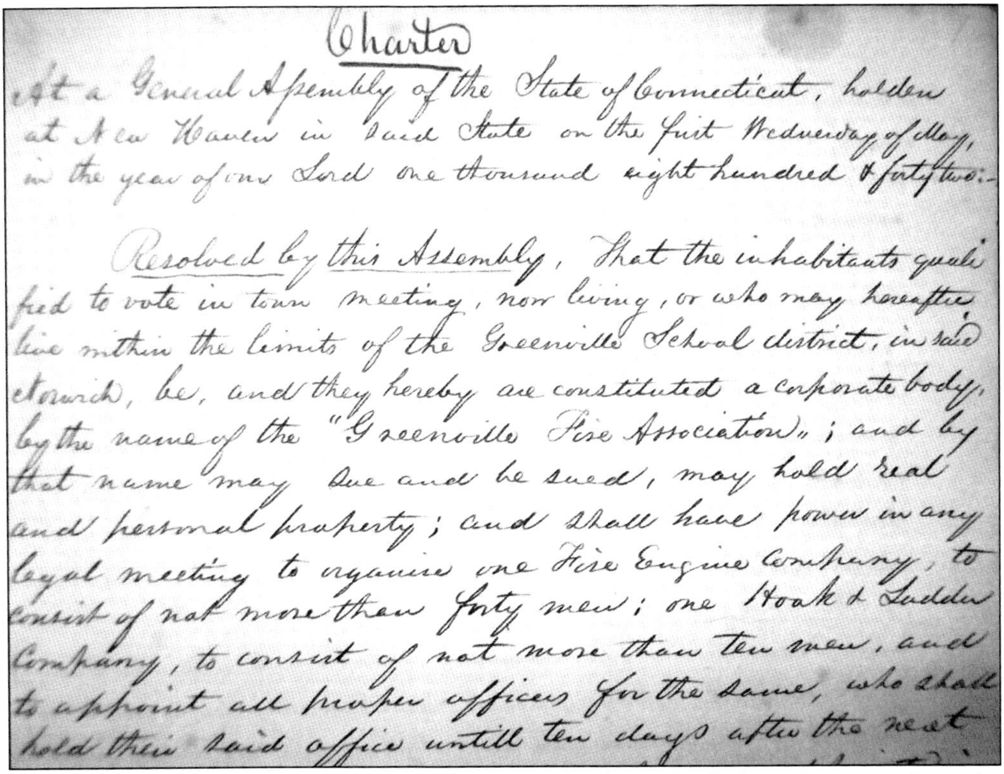

Charter

At a General Assembly of the State of Connecticut, holden at New Haven in said State on the first Wednesday of May, in the year of our Lord one thousand eight hundred & forty two:-

Resolved by this Assembly, That the inhabitants qualified to vote in town meeting, now living, or who may hereafter live within the limits of the Greenville School district, in said Norwich, be, and they hereby are constituted a corporate body, by the name of the "Greenville Fire Association"; and by that name may sue and be sued, may hold real and personal property; and shall have power in any legal meeting to organize one Fire Engine Company, to consist of not more than forty men; one Hook & Ladder Company, to consist of not more than ten men, and to appoint all proper officers for the same, who shall hold their said office untill ten days after the next

In 1842, the Greeneville district of Norwich operated independently from the city and saw the need for fire protection following a disastrous fire at the Shetucket Cotton Company. A group of residents/mill workers from the Shetucket Cotton Company organized the Greeneville Fire Association in that year. One engine company to consist of not more than 40 men and one hook and ladder company to consist of not more than 10 men were formally established in the Greeneville Fire Association's charter. It is uncertain if the hook and ladder company was actually organized in 1842.

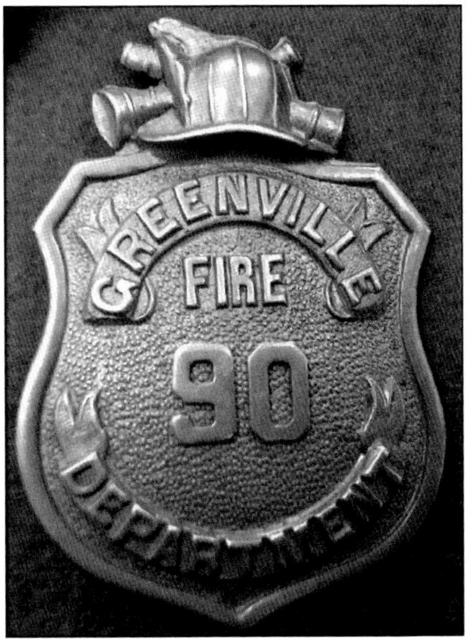

The Greeneville Fire Department was officially formed under Chief John Wilbur in 1862. The department existed for 15 years before being incorporated into the Norwich Fire Department in 1876 when the city annexed the village of Greeneville. Spelling variations of Greeneville have been noted on documents and badges, but the village's official spelling is Greeneville.

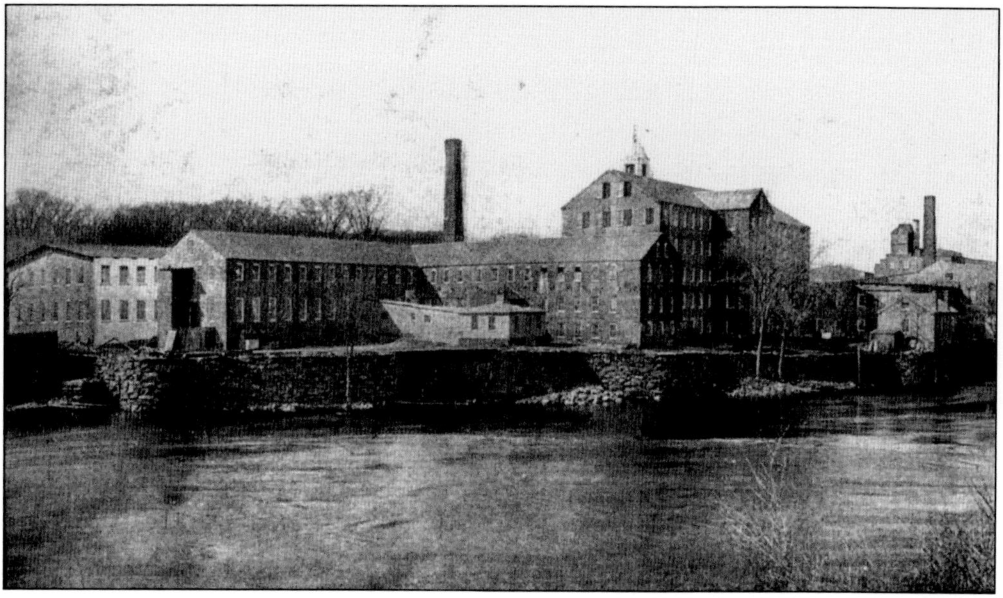

The *Quinebaug* hand engine was similar in form to the *Torrent* and was placed in service in 1846 in the Greeneville district of Norwich. The location of the apparatus in 1846 is undetermined but presumed to be in close proximity to the Shetucket Cotton Company (above). The Shetucket Cotton Company was founded by William P. Greene, for whom Greeneville was named. The *Quinebaug* hand engine was in service from 1846 to 1861 when its name was retired. In 1862, the apparatus was renamed Greeneville No. 1 and served for an additional five years at that title.

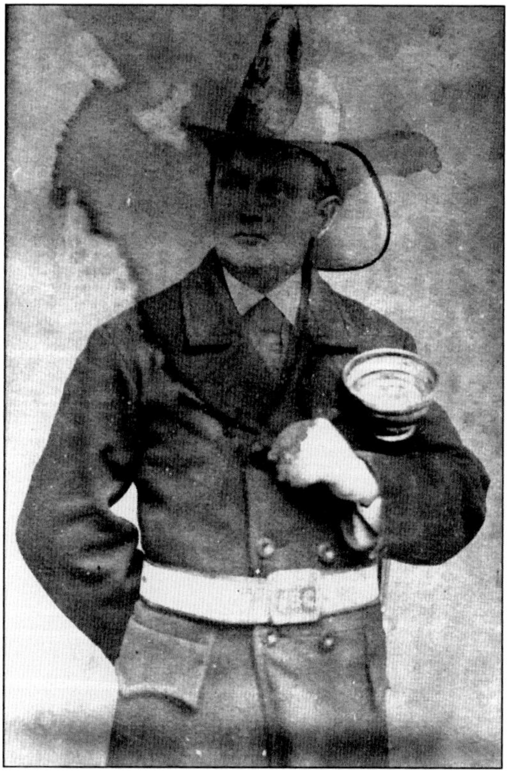

Joseph B. Carrier was appointed to the rank of chief in 1874. In 1876, he was made permanent fire chief and assumed command of both the Norwich and Greeneville Fire Departments. His duties were increased to fire marshal and superintendent of the fire alarm system, which had been installed earlier in 1876. Carrier's position of career chief laid the groundwork for Norwich Fire Department's permanent force, which began in 1902. Carrier served as chief of the Norwich Fire Department until his death in 1890.

Organized on June 15, 1888, the Greeneville Hook and Ladder Company operated out of a small wooden building located on North Main Street. The small wooden building that housed the apparatus was destroyed by fire sometime before 1896. In 1896, the Greeneville Hook and Ladder Company moved to 452 North Main Street and shared quarters with the Shetucket Steam Engine Company.

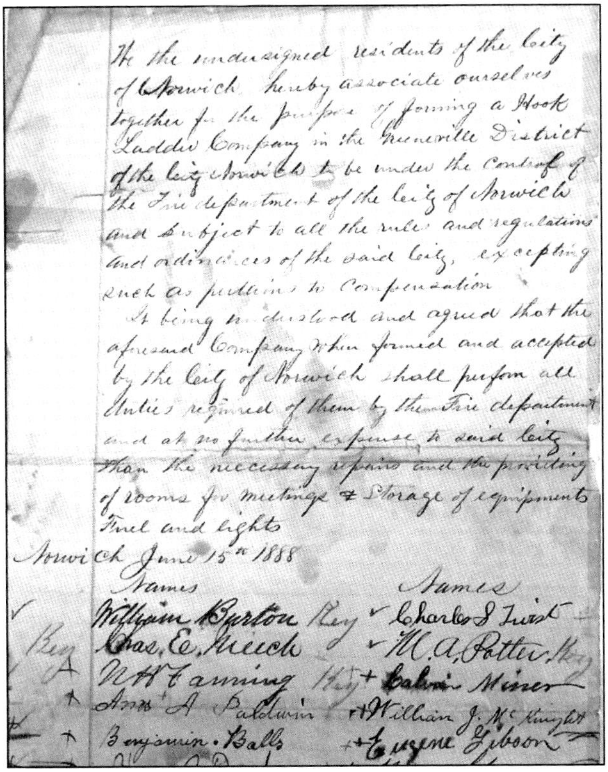

We the undersigned residents of the City of Norwich hereby associate ourselves together for the purpose of forming a Hook Ladder Company in the Greeneville District of the City Norwich to be under the control of the Fire department of the City of Norwich and subject to all the rules and regulations and ordinances of the said City, excepting such as pertains to Compensation

It being understood and agreed that the aforesaid Company when formed and accepted by the City of Norwich shall perform all duties required of them by the Fire department and at no further expense to said City than the necessary repairs and the providing of rooms for meetings & storage of equipments Fuel and lights

Norwich June 15th 1888

Names — Names

William Burton — Charles J. Twist

Chas. E. Meech — M. A. Potter

N H Canning — Calvin Miner

Amos A. Baldwin — William J. McKnight

Benjamin Balls — Eugene Gibson

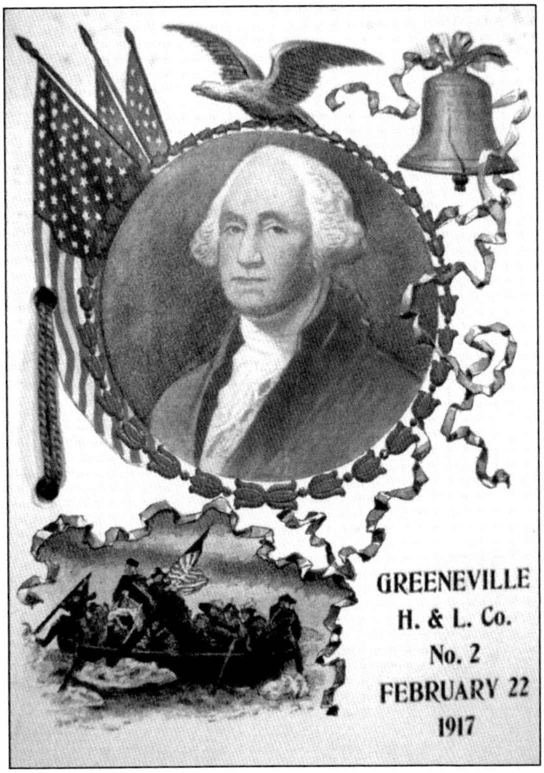

This 1917 event program is from the Greeneville Hook and Ladder Company. Volunteer fire companies of the 1800s and early 1900s were known for having elaborate annual gatherings in which other fire companies were invited. Greeneville's two fire companies remained volunteer until 1922 when permanent men were placed in the station.

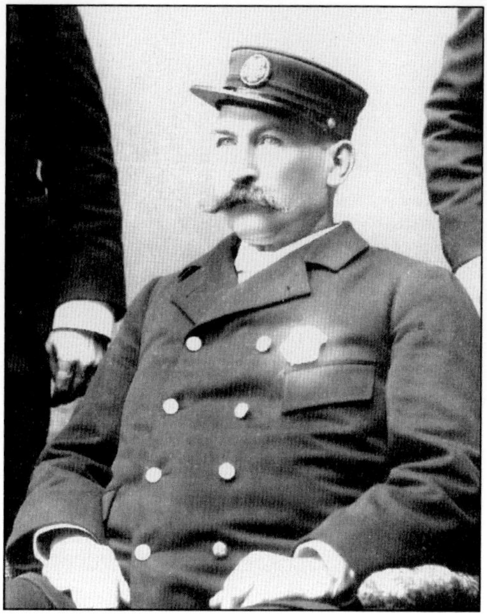

Chief Louis W. Greenberg served as fire chief following the death of Chief Joseph B. Carrier in 1890. Greenberg served in that capacity for 11 years before retiring in 1901. During his term in office, he oversaw the construction of a larger and more modern firehouse in the Greeneville district.

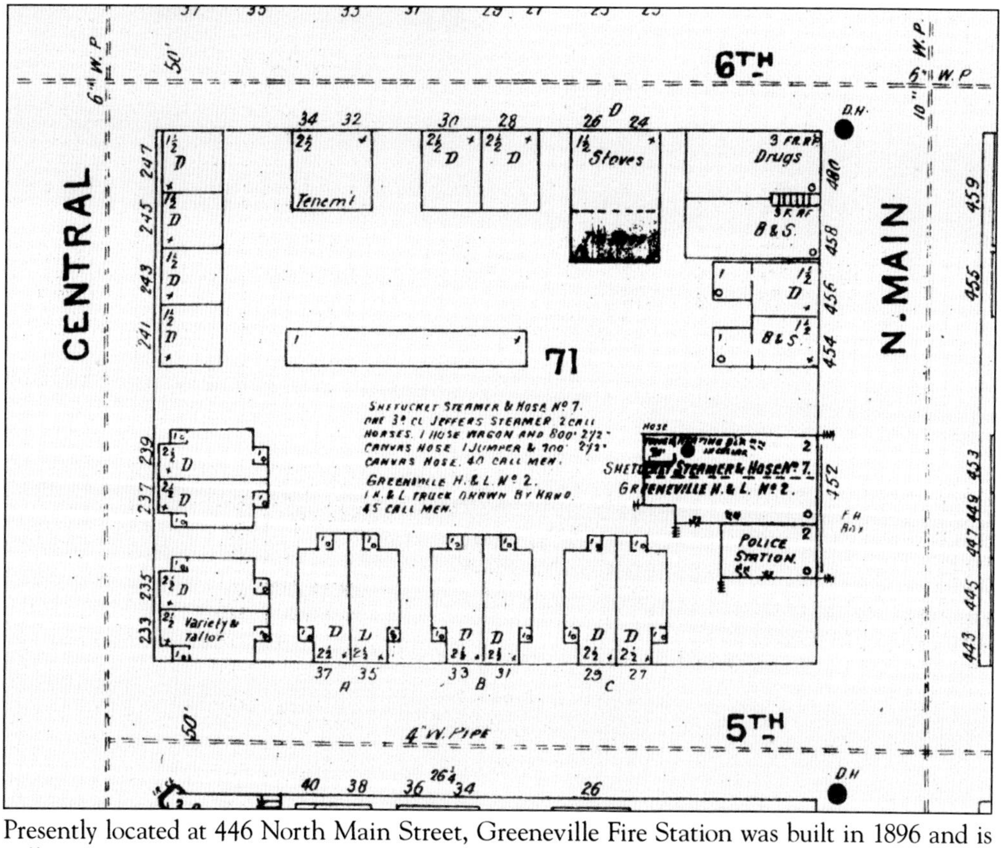

Presently located at 446 North Main Street, Greeneville Fire Station was built in 1896 and is still in operation today. The fire station was built to accommodate Shetucket Steam Engine Company No. 7 and the Greeneville Hook and Ladder Company No. 2, as well as a new substation for the Norwich Police Department.

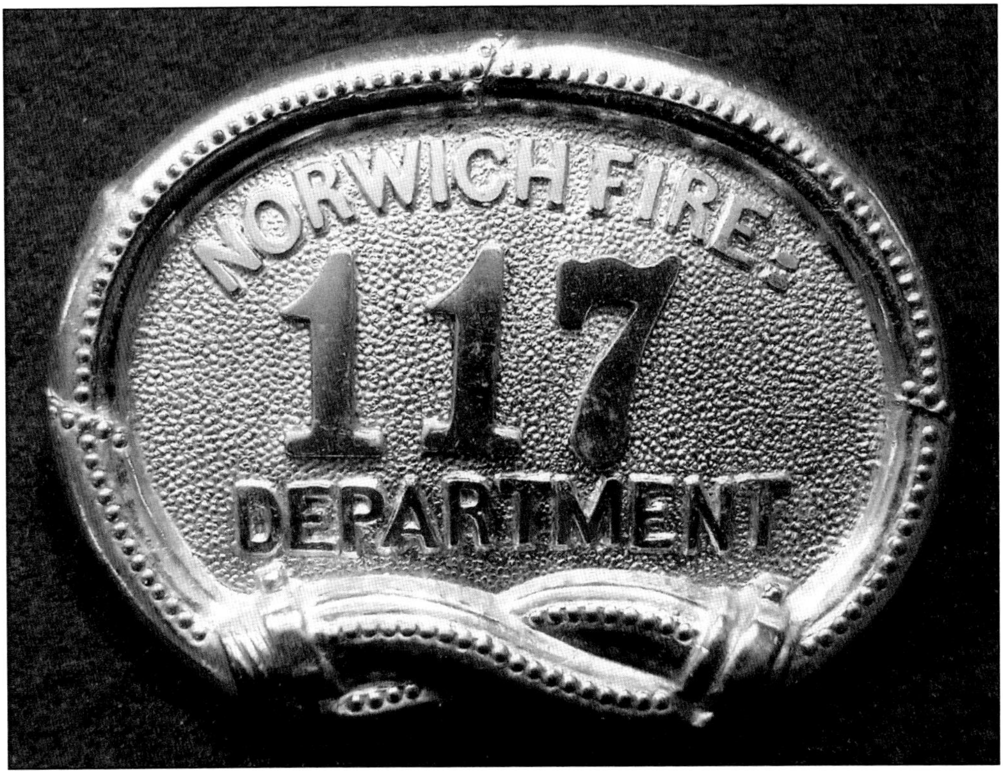

These Norwich badges are from various periods in the department's history. The oval badge is encompassed by a riveted leather fire hose, and the shield-style badge displays a multitude of fire department paraphernalia.

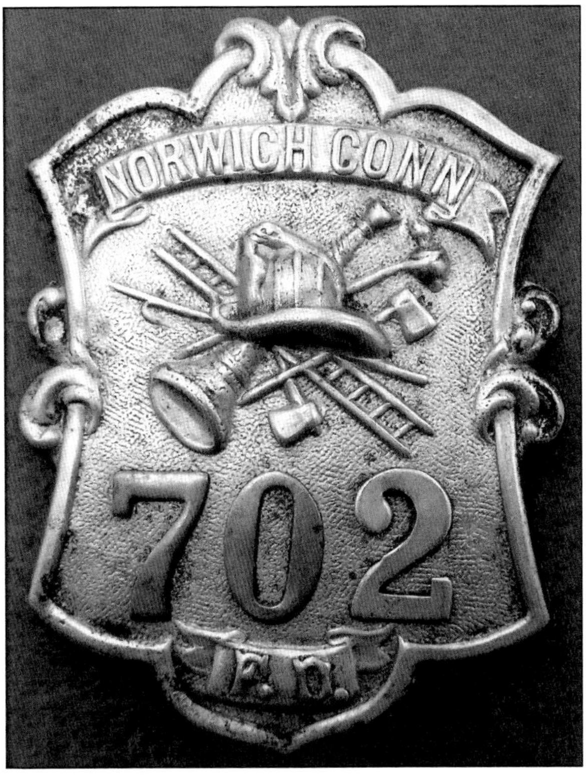

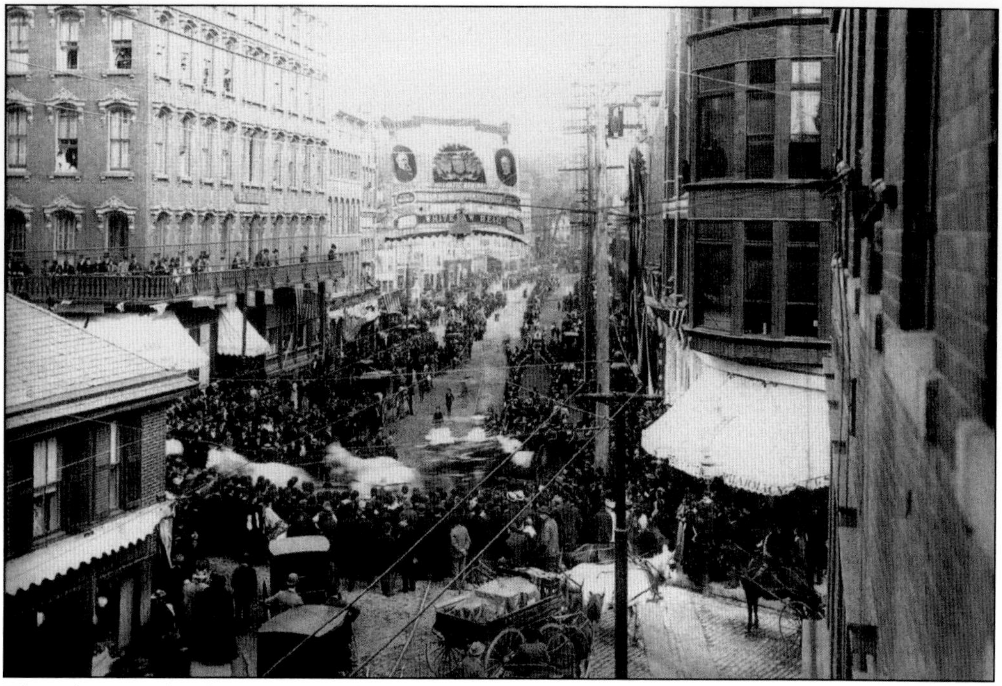

This photograph of a 1892 presidential campaign parade shows a Norwich Fire Department steam engine being pulled up lower Broadway. This intersection is referred to today as Courthouse Square.

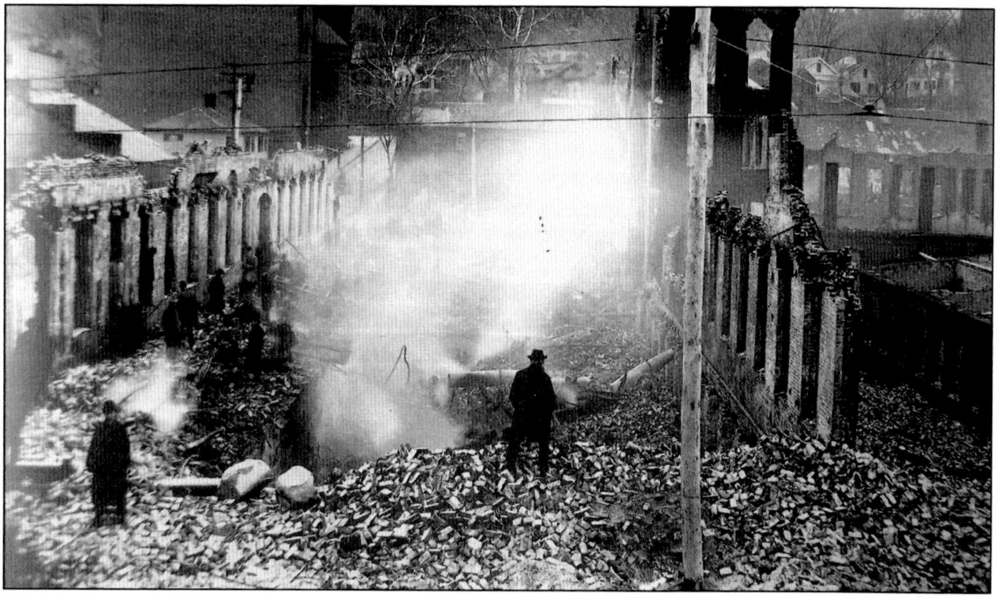

An explosion at the Hopkins and Allen firearms factory on the morning of February 4, 1900, jolted the citizens of Norwich awake. The large multistory brick mill was located between Chestnut and Franklin Streets and suffered major damage with the fire that ensued. The factory was instrumental in supplying the Union forces with rifles during the Civil War. The fire loss, totaling $175,000, stressed the need for a full-time fire service to respond immediately from its quarters.

Two

THE CAREER DEPARTMENT

Beginning in 1902 with a partially paid department, Norwich underwent great changes throughout the early 1900s before becoming a fully paid service in 1922. By utilizing the most modern technology of the time, the department was able to grow into a cutting-edge fire service.

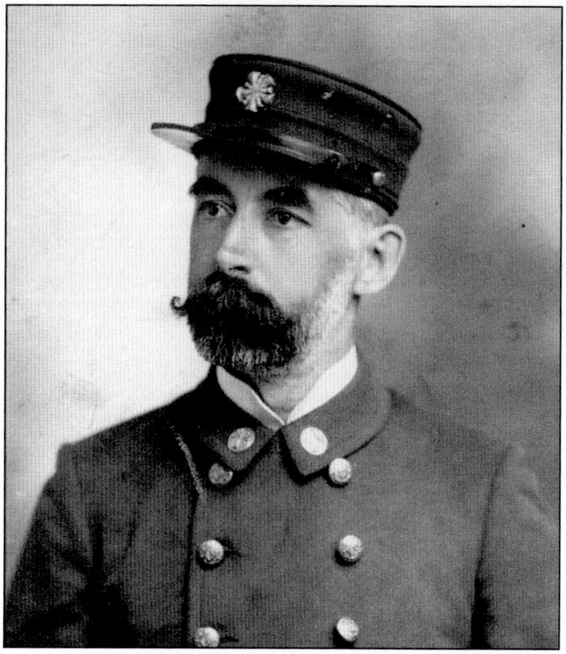

Chief Howard L. Stanton was elected chief engineer, fire marshal, and superintendent of the fire alarm telegraph in 1901. Stanton began his fire department career as a member of Neptune Steam Fire Engine Company No. 2 as a signal boy in 1869. He continued serving the community as a member until 1899 when he retired at the rank of first assistant chief. He was then hired as fire chief in 1901 and served until 1929 when he retired after 57 years as a Norwich fireman. Stanton was instrumental in forming the career fire department, which began in 1902.

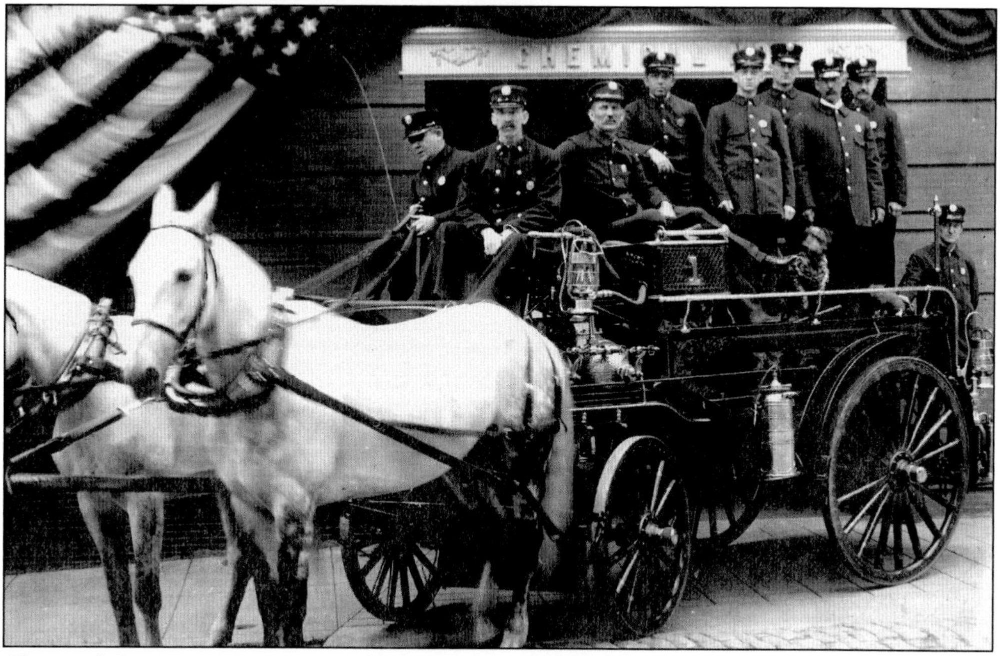

Chemical Company No. 1, originally located at 181 West Main Street in 1902, was relocated to 69 Main Street in 1907. The first career firemen of the Norwich Fire Department were Capt. Alvin Lewis, hoseman Patrick J. Caples, and driver Edward F. Stinson. This addition of permanent firefighters on November 1, 1902, began the transition from an all-volunteer department to the new career fire department that would soon follow. Chemical Company No. 1 was in existence from 1902 to 1921 when its call sign was retired. The photograph above shows Chemical Company No. 1 outside its quarters in 1909 with three full-time firemen and six bunkers or part time firemen in the bed of the wagon.

The men of Chemical Company No. 1 are pictured outside their firehouse located at 67–69 Main Street. The Chemical Company's captain is pictured third from the left. His double-breasted uniform and lapel pins signify his higher rank.

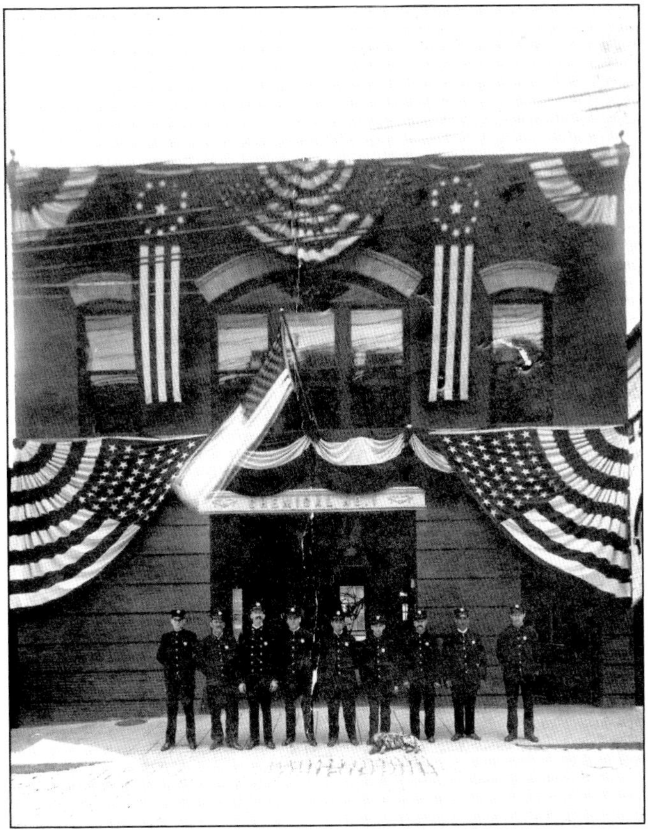

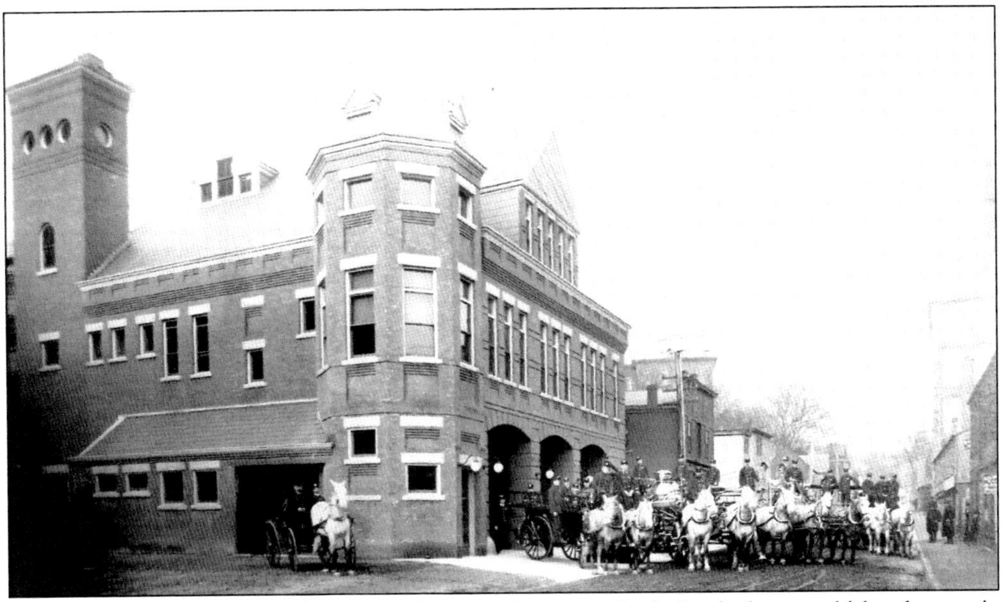

Completed in 1904, Central Fire Headquarters became the hub of what would be the newly formed career department. This beautiful Queen Anne–style firehouse was designed to house multiple fire companies, thereby consolidating the downtown fire district. This commanding building would be in service for 94 years.

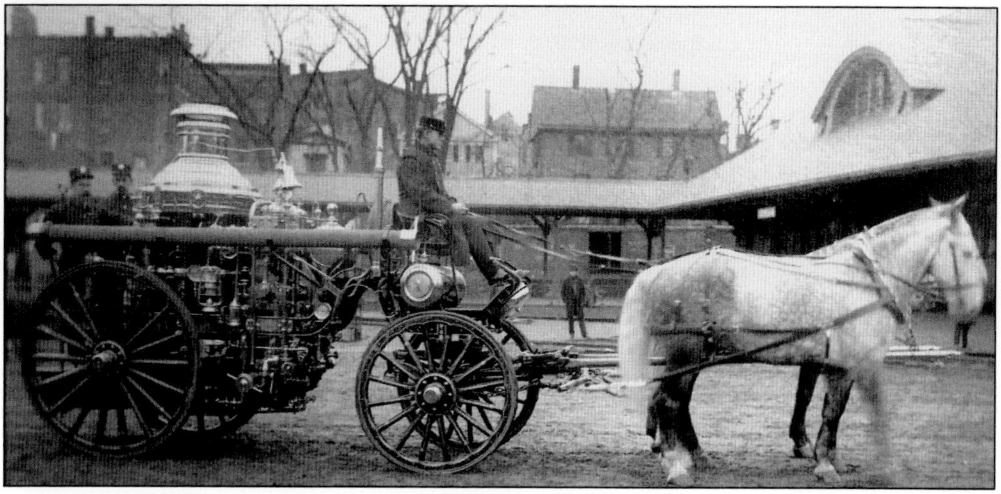

Located at headquarters on Chestnut Street, Engine Company No. 1 was organized in 1905. The 1905 American LaFrance Metropolitan steam engine was known as a third-class steam engine capable of 600 gallons per minute. The apparatus weighed 6,800 pounds and was outfitted with a draft pole for either a two- or three-horse hitch.

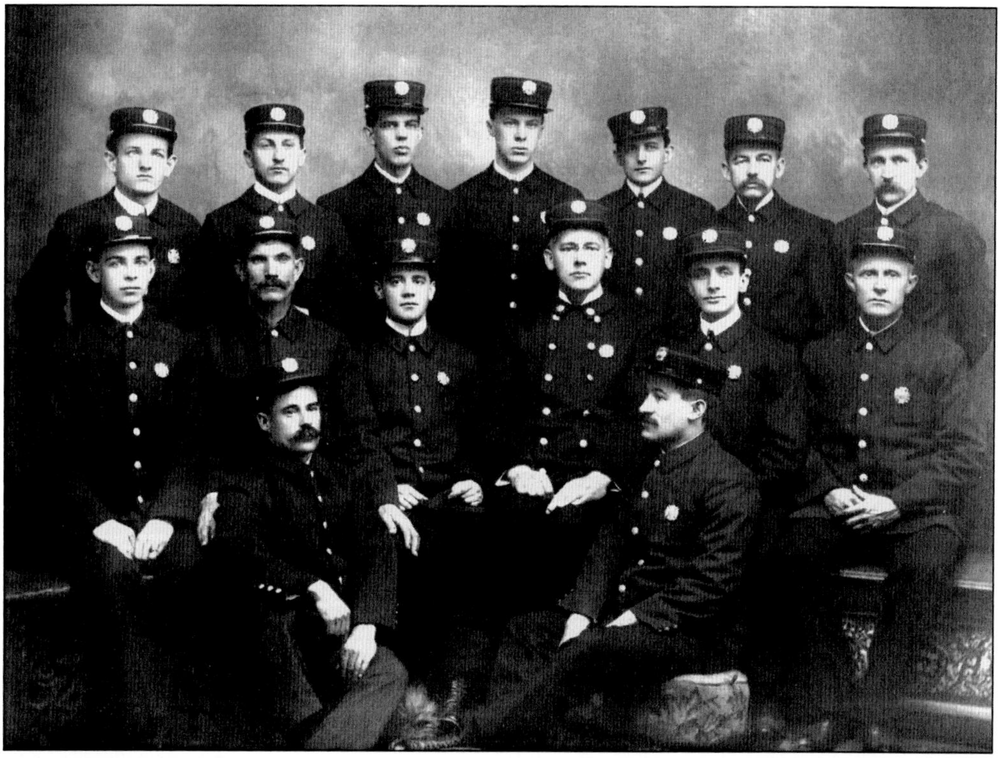

Assembled in this 1905 picture are the original members of Engine Company No. 1. From left to right are (first row) engineer ? Williams and hydrant man ? Hors; (second row) hoseman ? Young, driver ? McCaffrey, chaplain ? Carey, Capt. ? Brown, pipeman ? Kind, and hoseman ? Tuttle; (third row) hoseman ? Arliuau, pipeman ? Beiag, pipeman ? Brown, pipeman ? Hough, hoseman ? Ferguson, hydrant man ? Smith, and hoseman ? Hopkins.

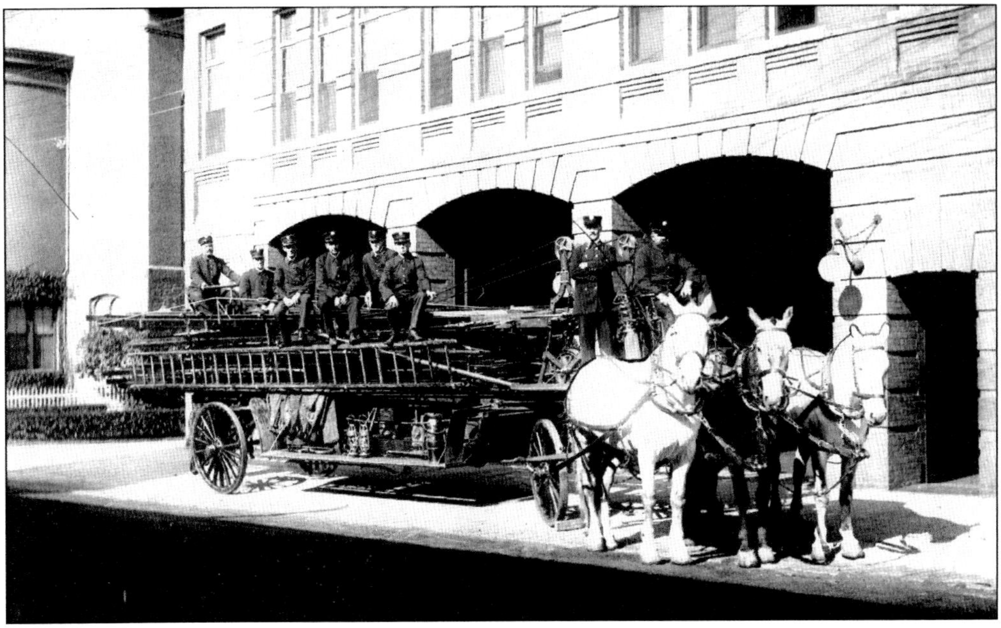

Aerial truck No. 1 was purchased by the city prior to the completion of fire headquarters in 1904. The ladder truck was stored in a nearby barn until the newly formed career department could move the apparatus into the new firehouse. Some years later, the apparatus was replaced with a more modern apparatus and the 65-foot Dietz ladder was relocated to the Greeneville section of Norwich and became truck company No. 2.

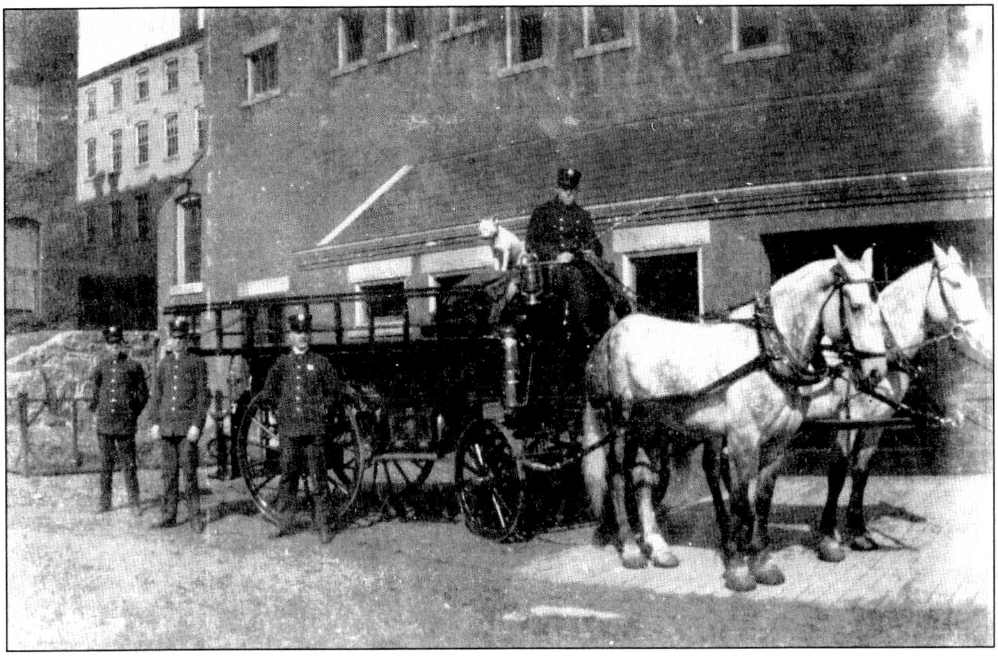

The Norwich Fire Department's hose company No. 5 was the third apparatus stationed at headquarters. This new hose wagon, purchased in April 1903, was placed in service temporarily in the Niagara hose station on Broadway before being moved to headquarters on Chestnut Street.

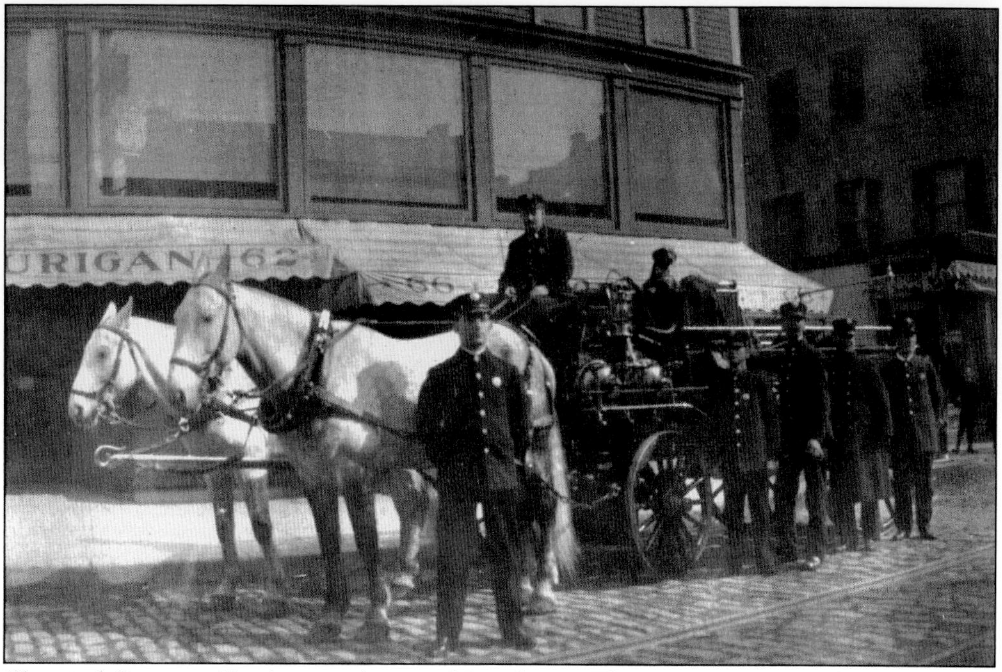

Chemical Company No. 1 is seen here on Main Street. The apparatus was outfitted with twin copper vessels that contained water and bicarbonate. By manipulating the valves on the vessels, pressurized water could be produced.

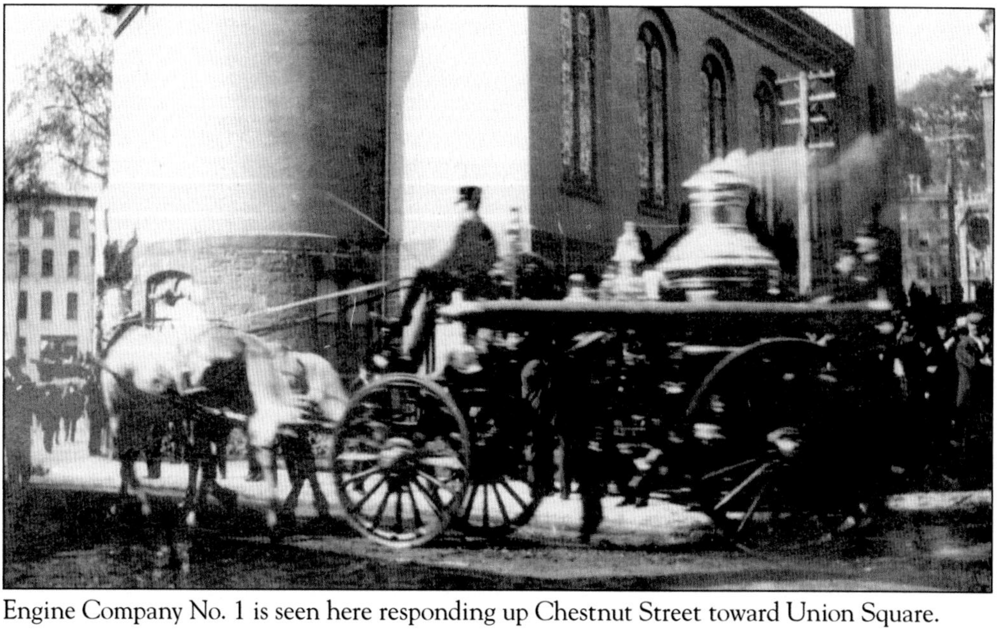

Engine Company No. 1 is seen here responding up Chestnut Street toward Union Square.

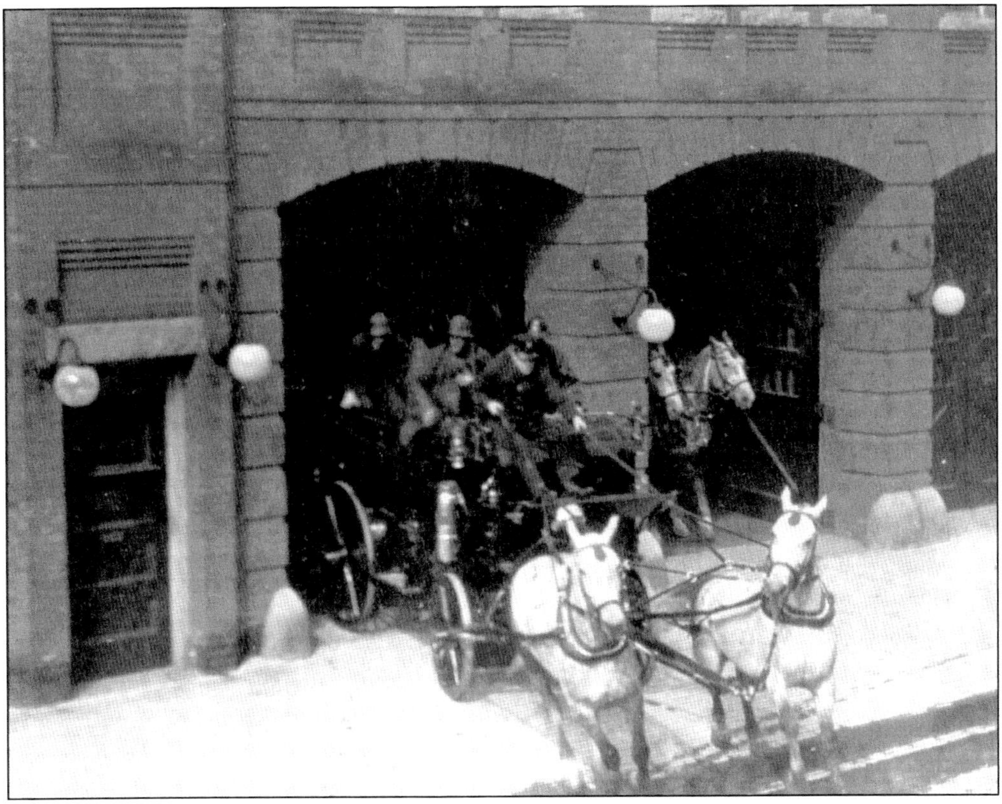

Hose Company No. 5 is seen here exiting fire headquarters with the horses of Engine Company No. 1 just edging out of the adjacent apparatus bay.

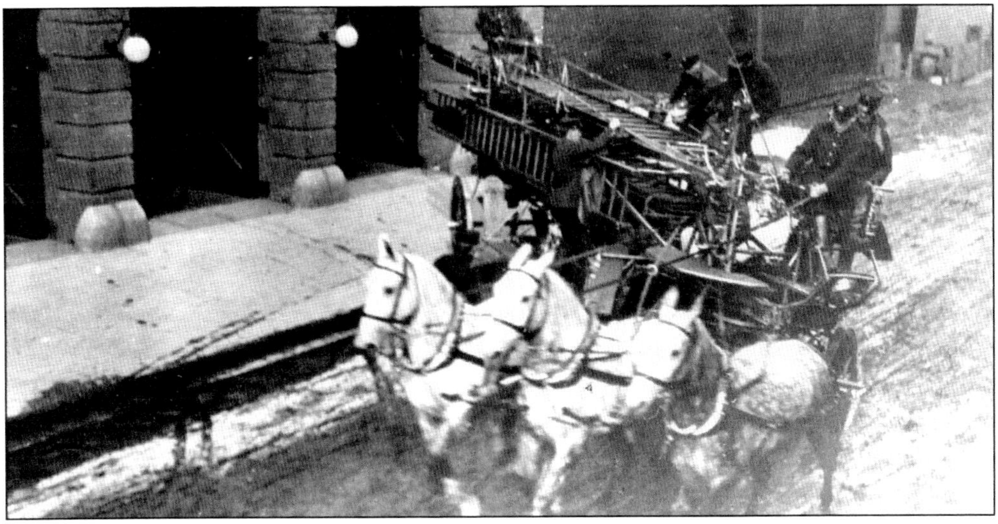

The aerial truck was of a tiller configuration. The front wheels were steered by the driver and horses while the rear wheels were controlled by the steering wheel of the tillerman. Apparatus of this style were seen in many cities throughout the United States because of their ability to negotiate tight corners.

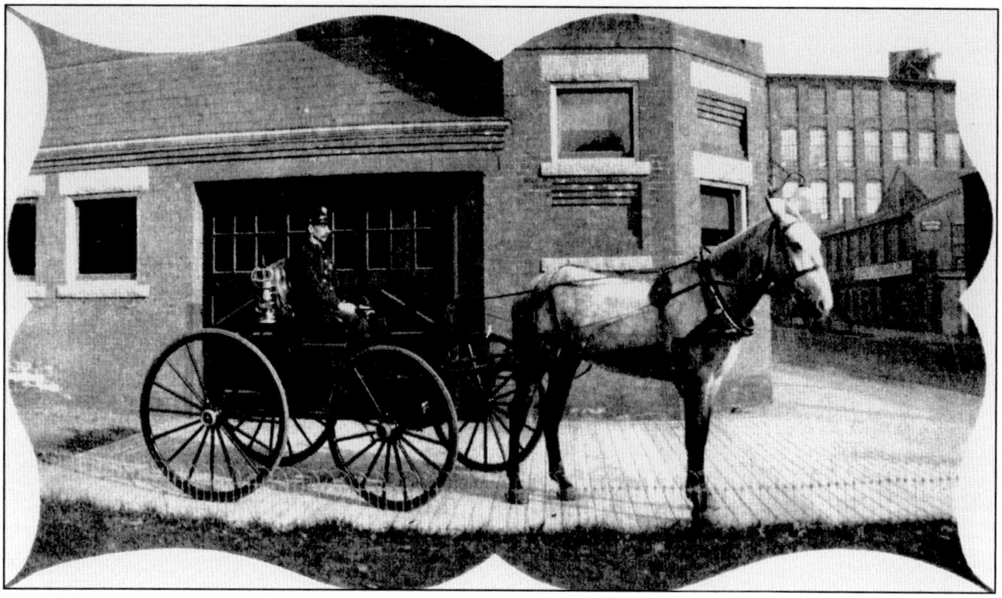

Chief Howard L. Stanton was able to drive or be chauffeured to calls ahead of the other responding units to organize the fire ground. The chief's apparatus bay was located on the side of the firehouse and contained a separate fire pole to allow the chief to slide from his office located in the turret of the second floor. Below, Chief Stanton is driven to an alarm with the remainder of headquarters following behind.

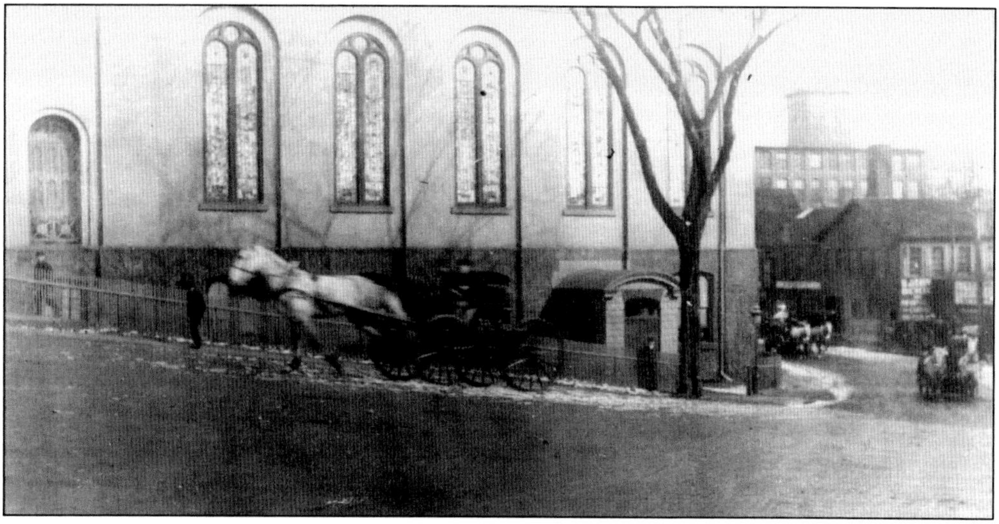

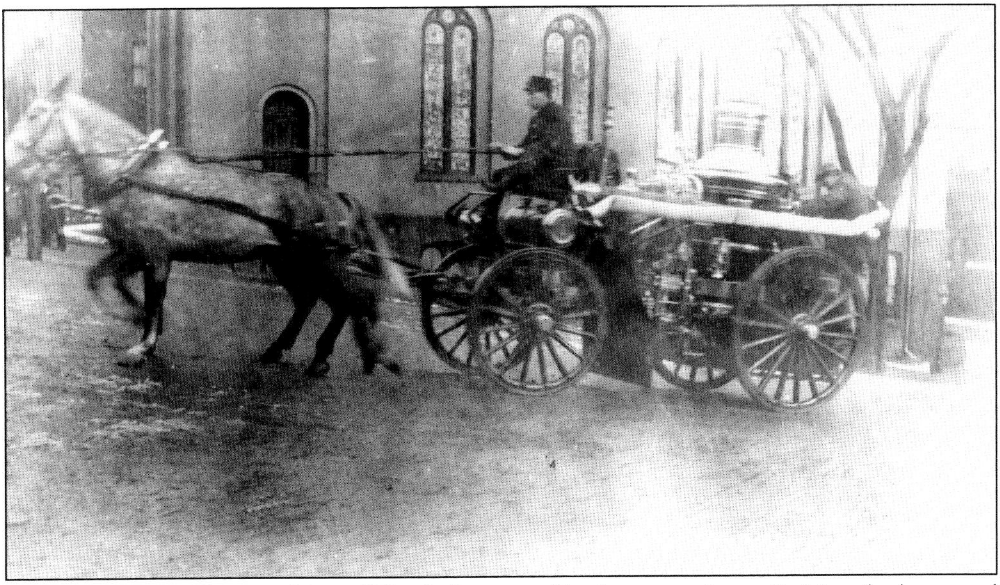

Engine Company No. 1 and Aerial Truck Company No. 1 are seen here climbing the lower end of Chestnut Street on their way into Union Square. The engine was configured with a two-horse hitch, while the heavier aerial truck was configured for a three-horse hitch.

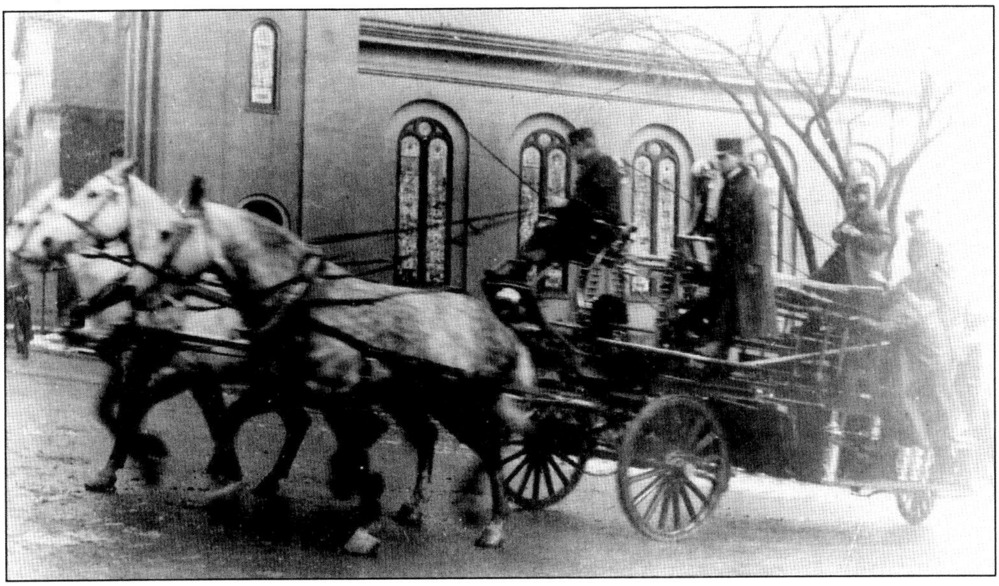

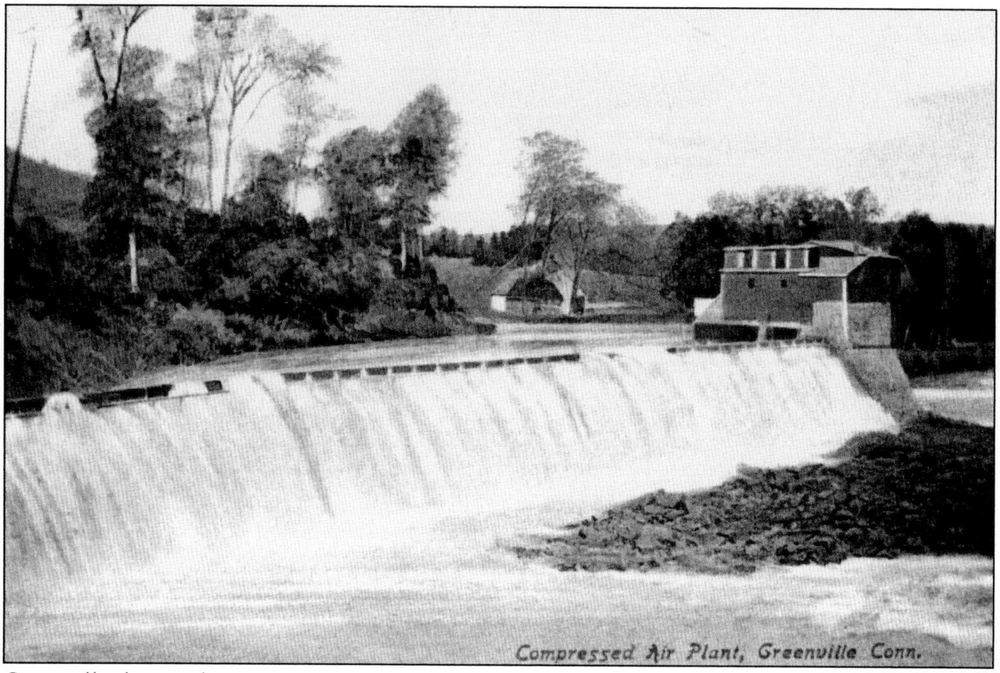

Originally designed to run small machinery in the downtown mills, compressed air lines were exploited by Chief Howard L. Stanton in 1905. The compressed air plant, located on the Shetucket River, produced pressurized air that was piped downtown. Stanton, a former machinist, drafted plans for a compressed air attachment that would allow for the engine company to instantly operate on the fire ground using compressed air. This time-saving step allowed for quicker application of water on the fire. The invention did work; however, no documentation has been discovered to suggest that it was fire tested. Below, engineer Calvin C. Williams (right) and an unidentified fireman operate engine No. 1 off compressed air outside headquarters.

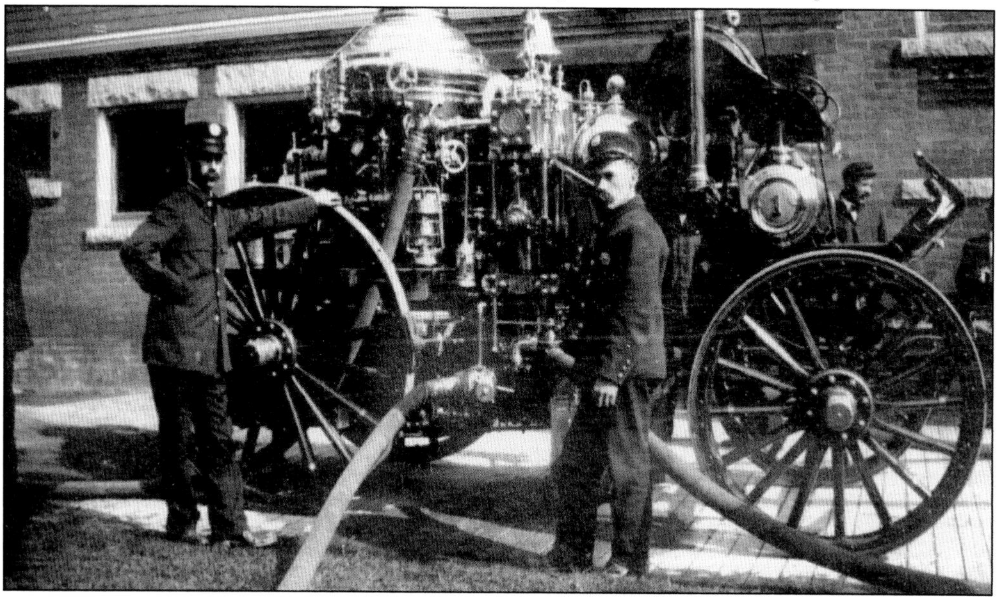

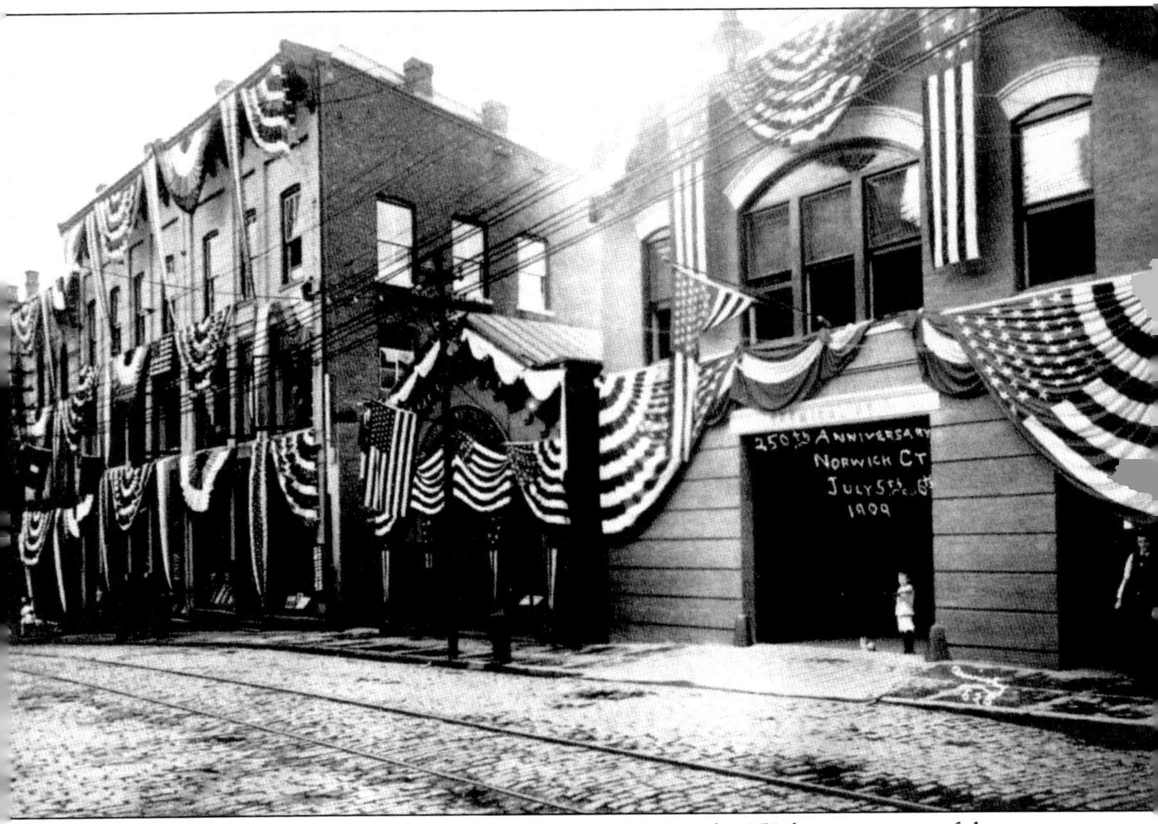

Chemical Company No. 1's station is shown here in 1909 on the 250th anniversary of the town. Most of the downtown buildings were draped in red, white, and blue for this festive occasion.

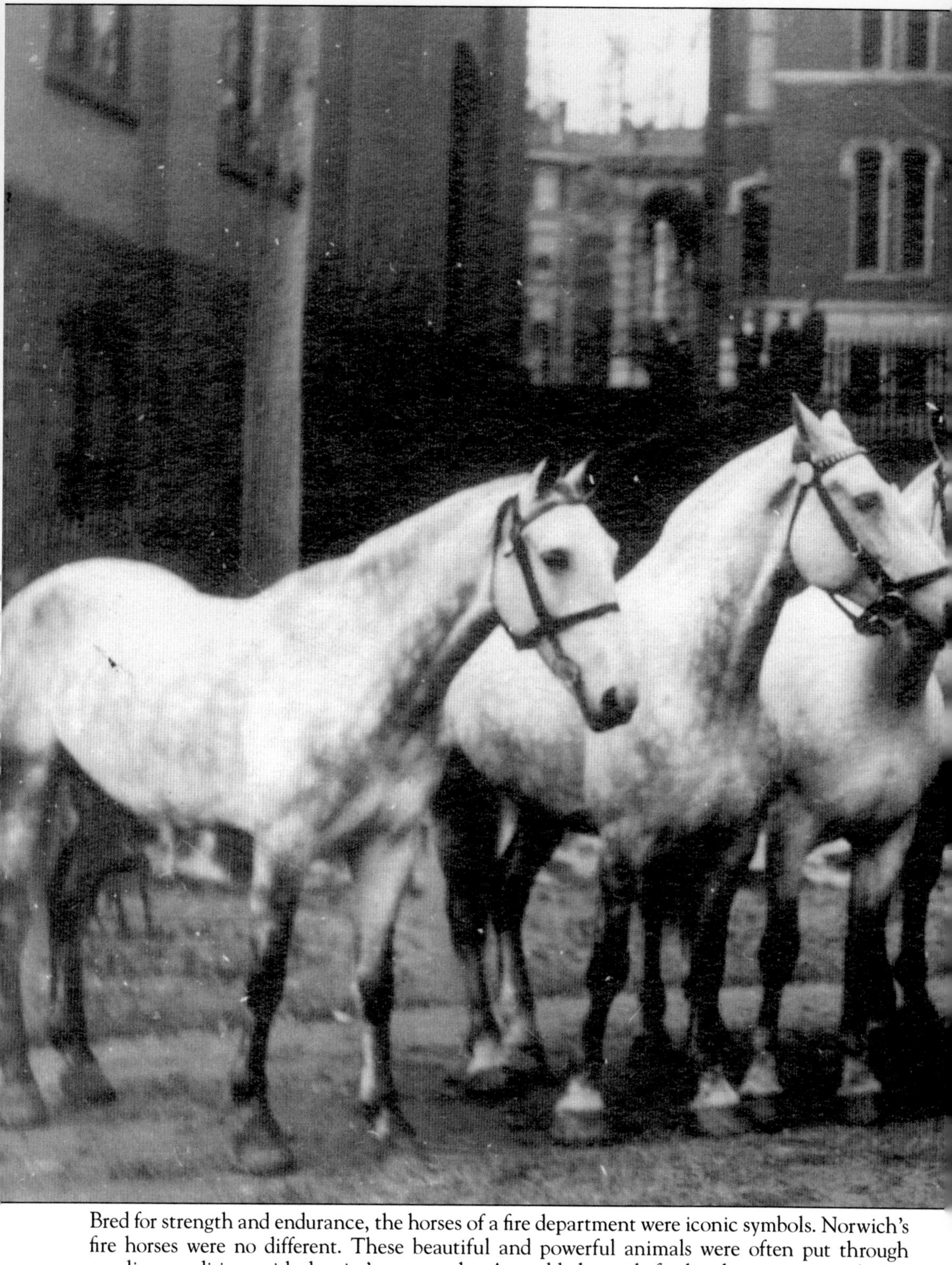

Bred for strength and endurance, the horses of a fire department were iconic symbols. Norwich's fire horses were no different. These beautiful and powerful animals were often put through grueling conditions with the city's topography. Assembled outside fire headquarters around 1905

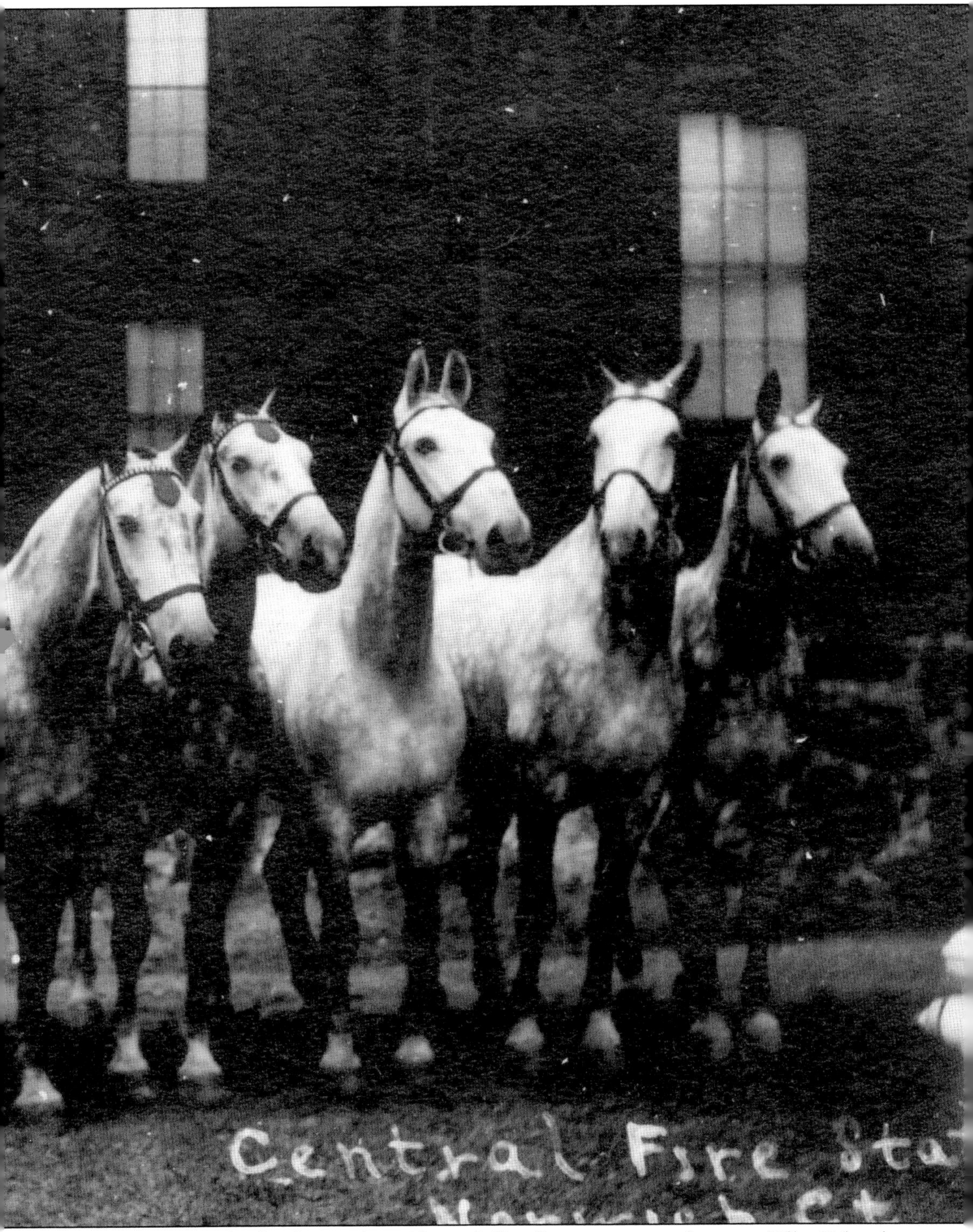

are eight of Norwich's hardest workers. Norwich Fire Department was home to Billy, Colonel, Charlie, Dave, General, Ginger, Jeff, Jennie, Jim, Major, Ned, Prince, Prize, Remy, Romy, Tamo, and Teddy.

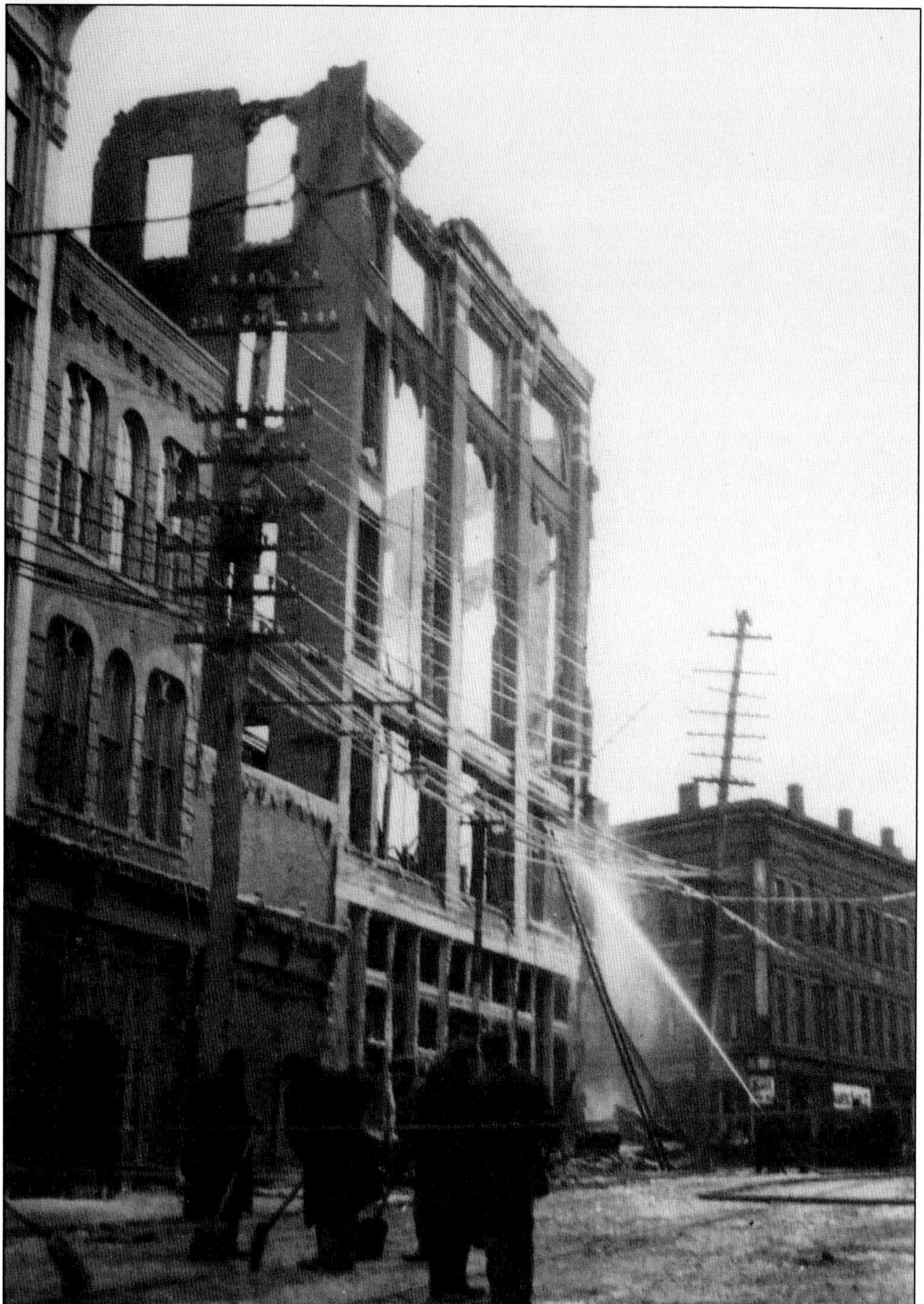

On February 9, 1909, a large fire swept through the Shannon Block at 2:30 a.m. This immense building, located on the corner of Shetucket and Main Streets, proved to be a daunting task for the emerging career fire department. More manpower was needed to prevent a conflagration of this size.

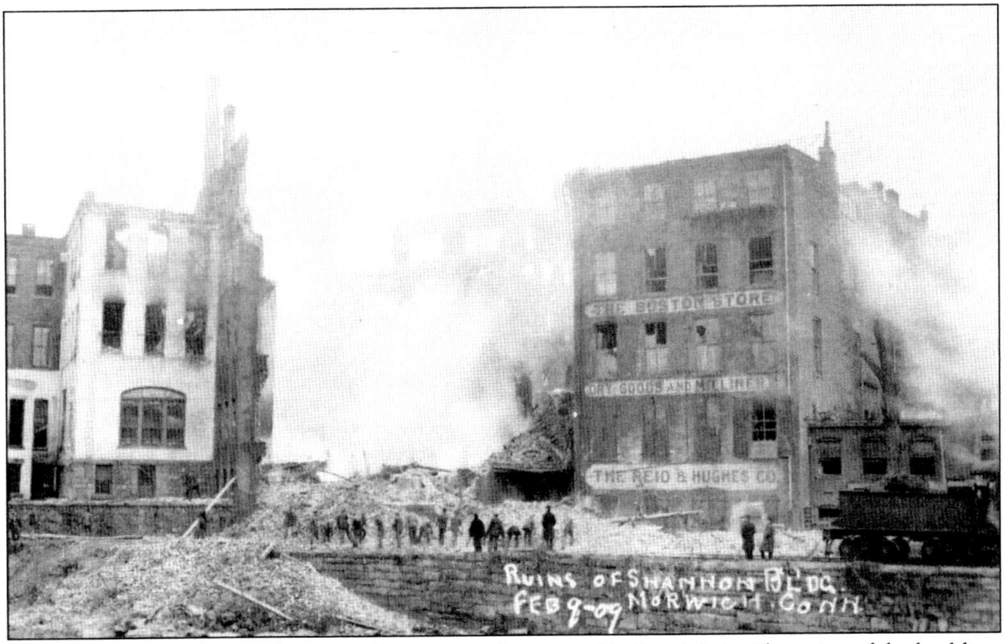

Firefighters worked throughout the day wetting down smoldering debris. The ruins of the building partially collapsed over the Norwich-Worcester Railroad tracks that ran behind the building.

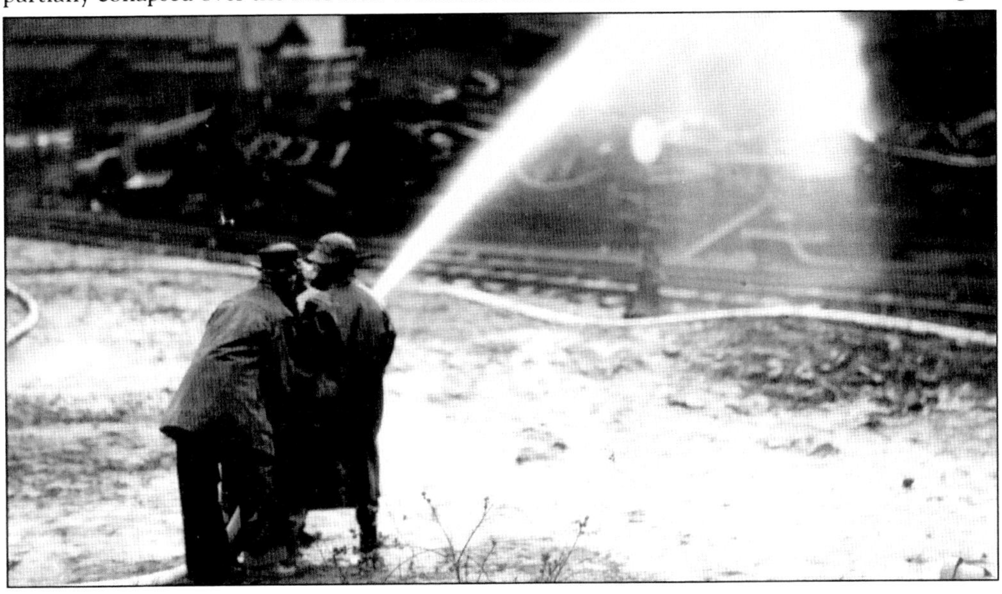

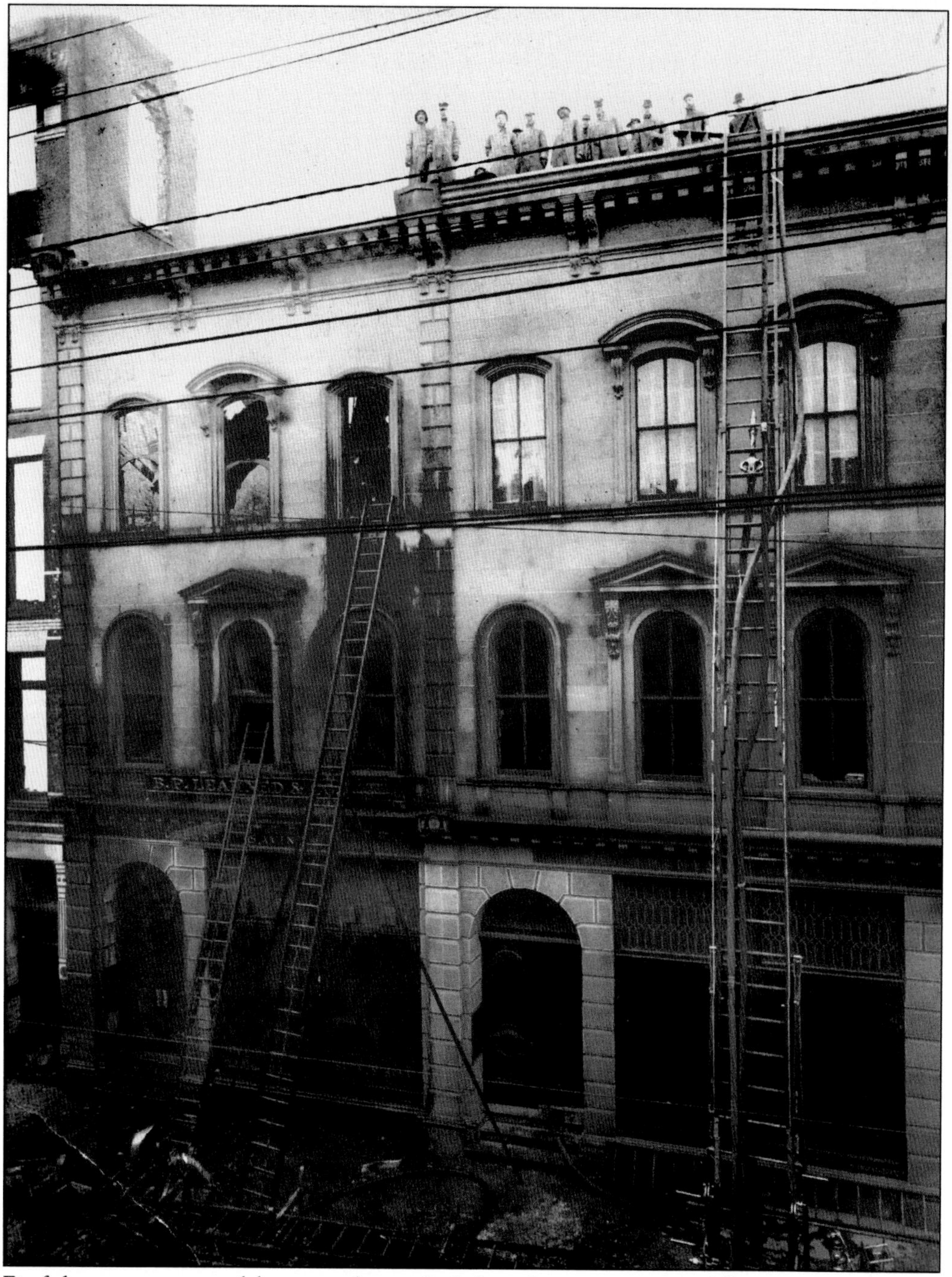

Firefighters are captured here on the roof of the adjacent structure following the Shannon Building fire. The surviving buildings in this photograph are located on Shetucket Street, known today as Chelsea Harbor Drive.

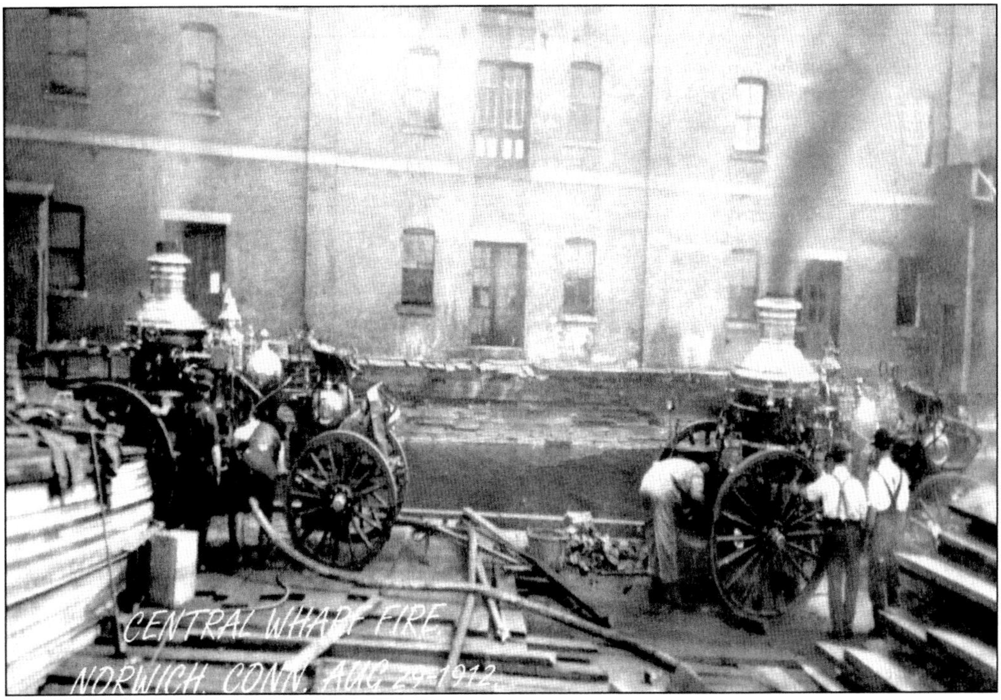

Ignited by vagrants, the large brick buildings of the Central Wharf were consumed by fire on August 29, 1912. Norwich Fire Department steam engines utilized water from the Thames River to combat the fire. The present-day marina and related grounds occupy the area affected by the fire.

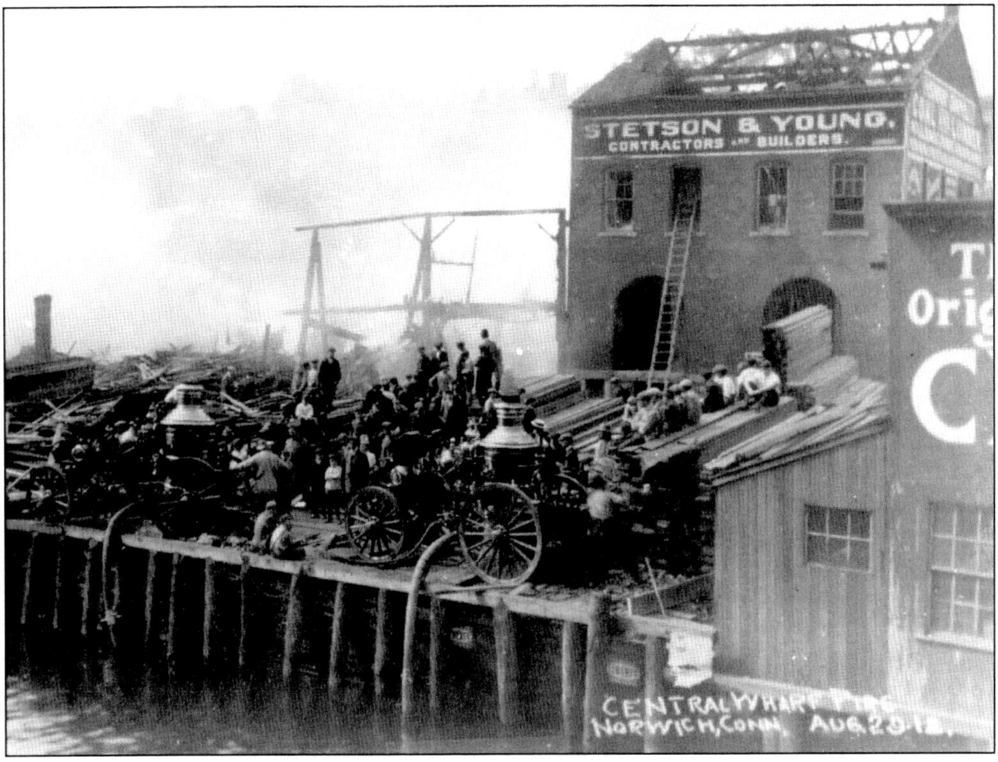

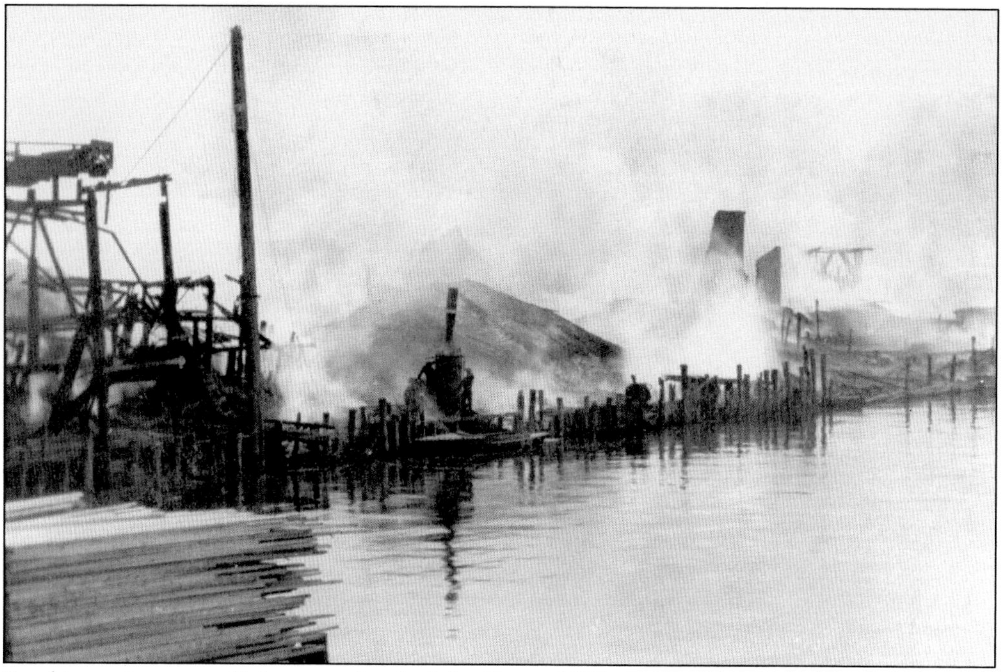

Smoking ruins were all that remained of the Central Wharf. In addition to large warehouses, the wharf also contained lumber and coal. The photograph above shows mounds of steaming coal among the other debris.

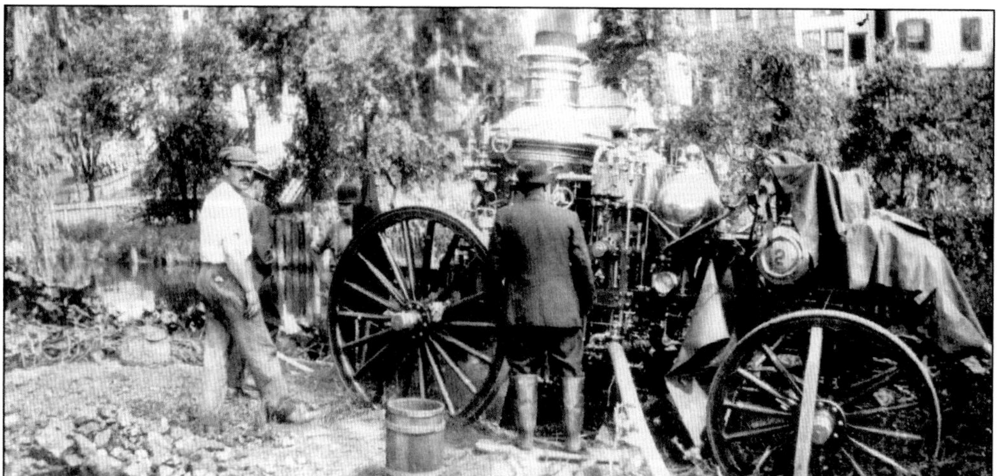

Engine No. 2 is pictured drafting water from the Yantic River during the Central Wharf fire of 1912. Engine No. 2, a 1907 American LaFrance Metropolitan steam engine, was stationed in Greeneville.

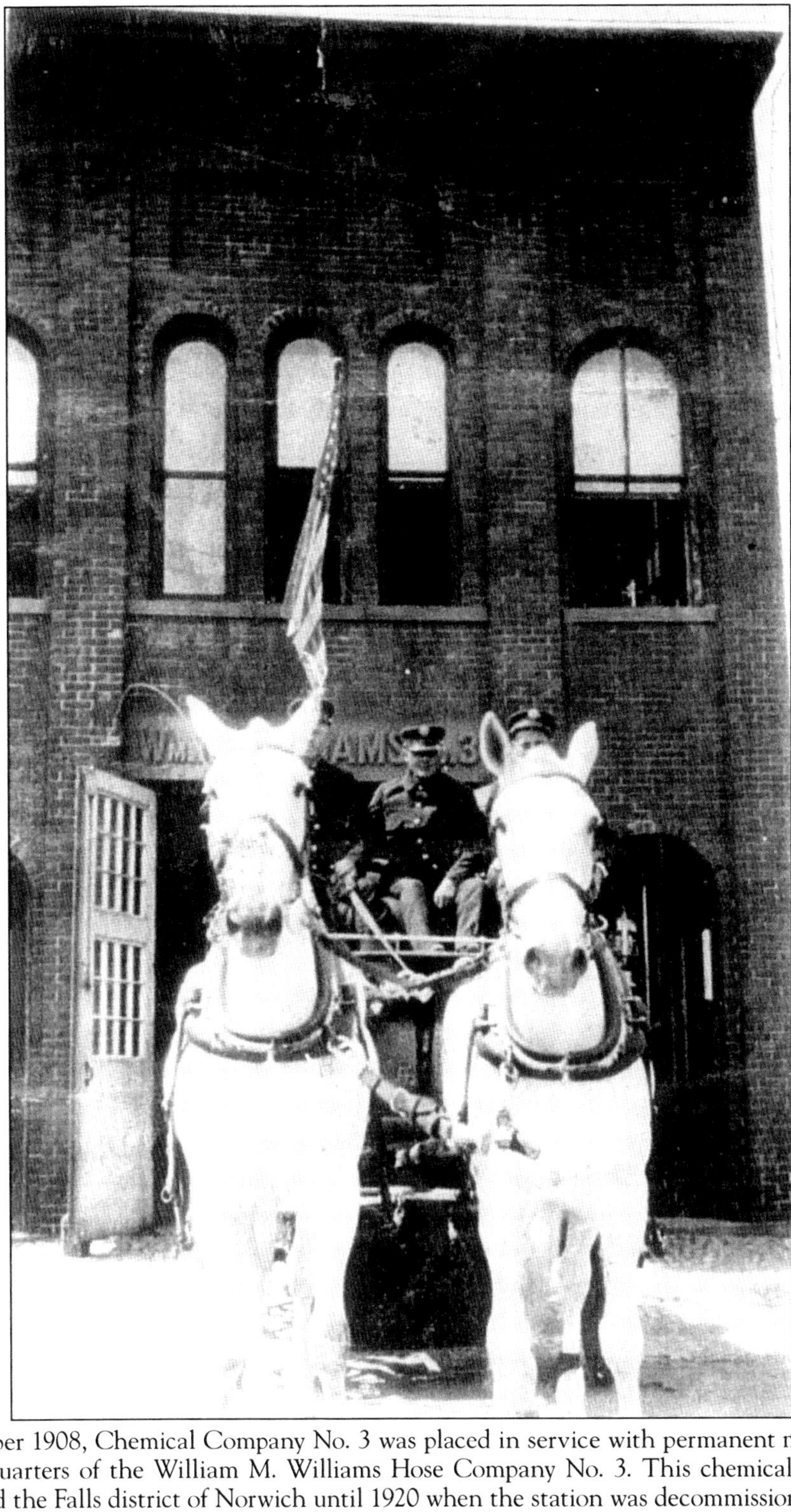

In October 1908, Chemical Company No. 3 was placed in service with permanent men at the former quarters of the William M. Williams Hose Company No. 3. This chemical company protected the Falls district of Norwich until 1920 when the station was decommissioned due to its age.

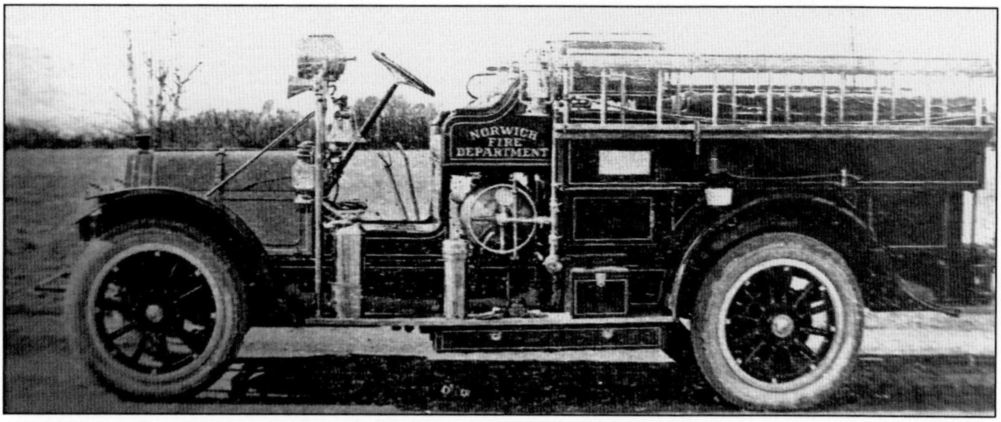

On August 11, 1910, the Norwich Fire Department utilized a rented automobile fire engine to conduct cost analysis and fire protection of a motorized apparatus versus the horse-drawn steamers of the time. Chief Howard L. Stanton conducted a time trial using a simulated fire at the corner of Mount Pleasant and Elizabeth Streets. Of the three engines that responded, one was an automobile from headquarters (one and a quarter miles away), the second was a horse-drawn steamer from station No. 5 (half a mile away), and the third was a horse-drawn steamer from station No. 3 (one and three-quarters miles away). The results indicated that the automobile responding from one and a quarter miles away reached the fire one and a half minutes earlier than the nearest steamer, which was only half a mile from the fire. This demonstration showed the need for current technology and Stanton, along with the board of fire commissioners, began looking into manufacturers of automobile fire apparatus. After consulting Seagrave, American LaFrance, Knox, and Pope-Hartford, the Norwich Fire Department chose the nearby Pope-Hartford manufacturer to build its first motorized apparatus.

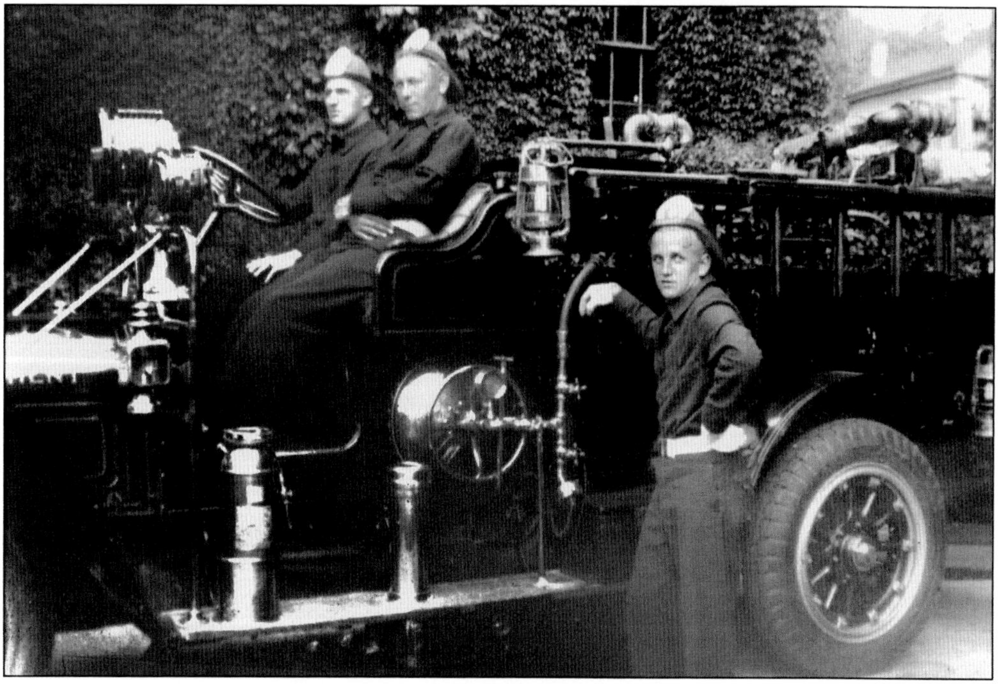

Three call men are pictured here on the Pope-Hartford apparatus in this 1934 photograph. The men were involved in a Fourth of July parade held by the city.

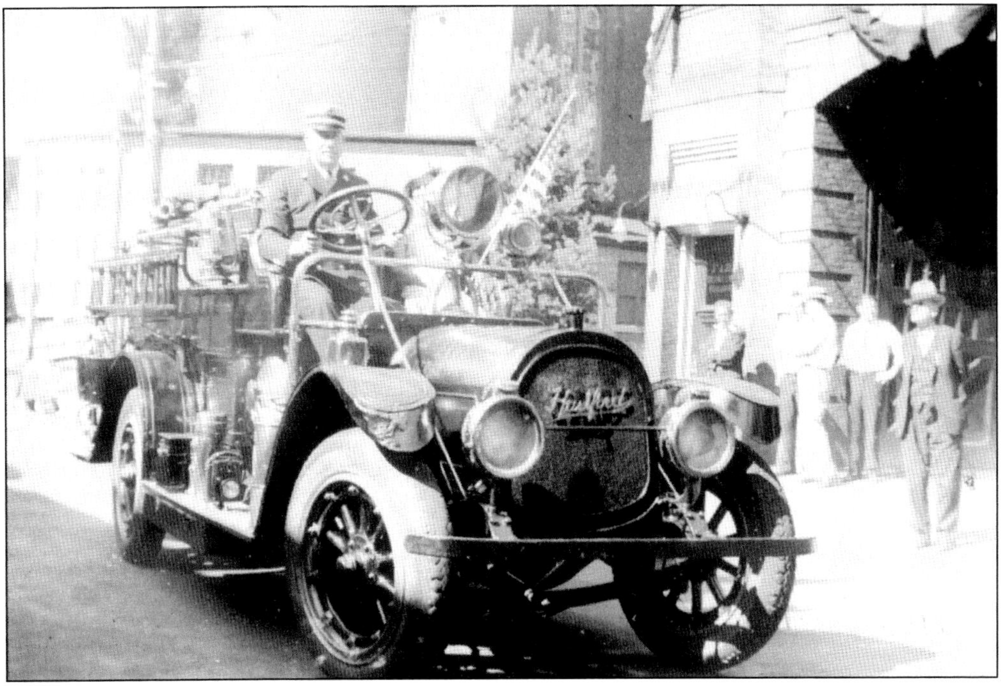

In this 1934 photograph, Chief Henry R. Taft is pictured at the wheel of the 1912 Pope-Hartford. Taft succeeded Howard L. Stanton in 1929 and served for 10 years before retiring.

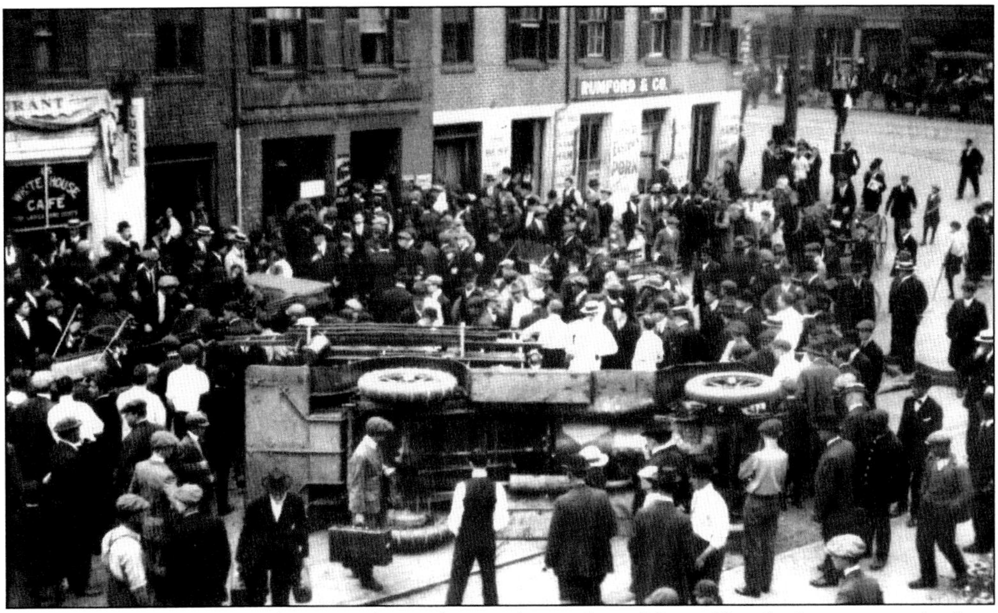

On May 22, 1914, an accident involving the newly purchased Pope-Hartford engine took place in Franklin Square. The fire apparatus suffered damages and was shipped back to Hartford for repairs at the factory.

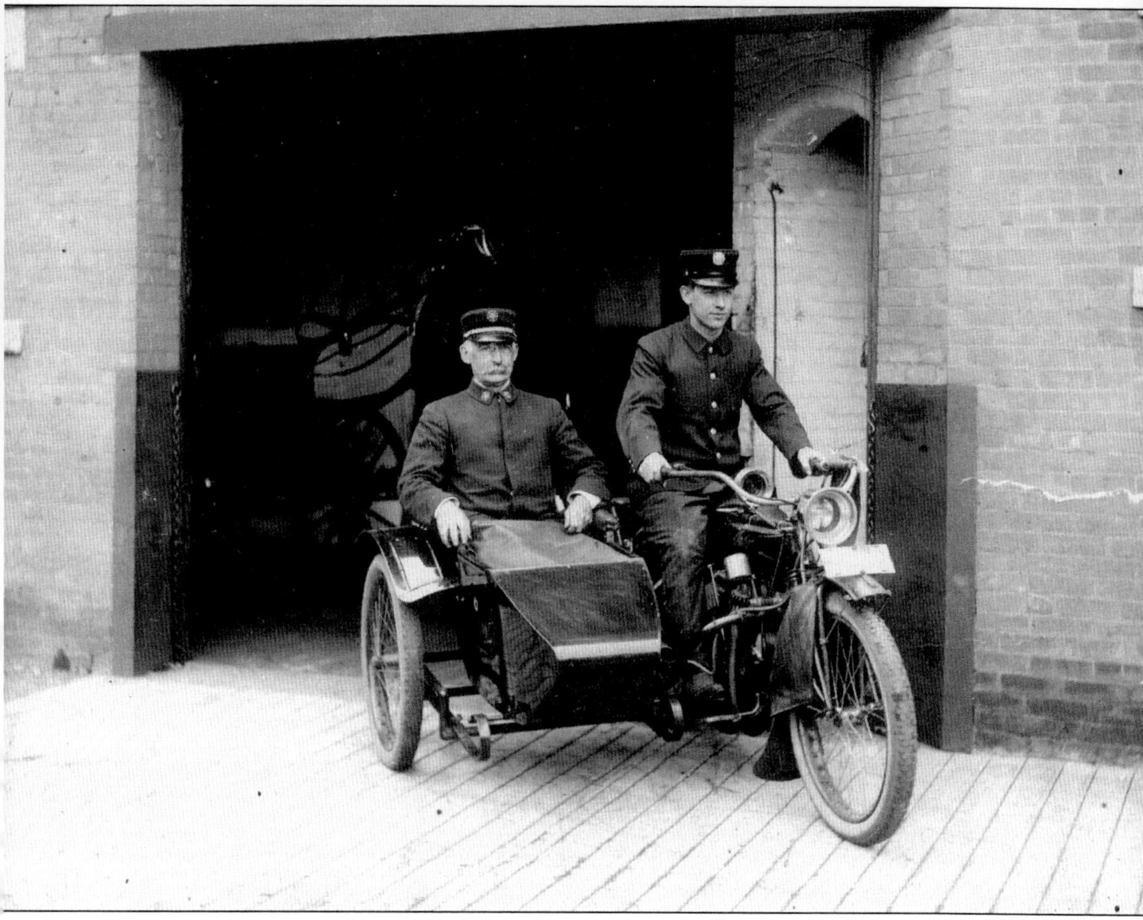

Chief Howard L. Stanton and his chauffeur pose on an Indian motorcycle outside fire headquarters around 1915. This advancement in technology allowed the fire chief to be transported quickly to all areas of the fire district.

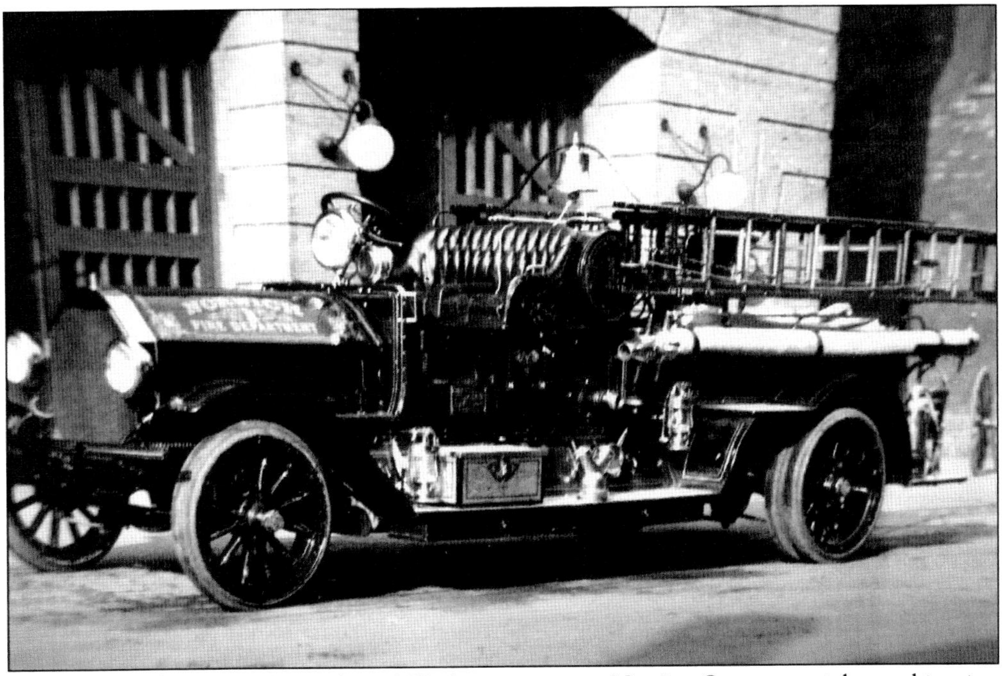

Purchased for $9,000 in December 1915, the new engine No. 1, a Seagrave triple-combination chemical, pump, and hose car, was placed in service at the Chestnut Street headquarters.

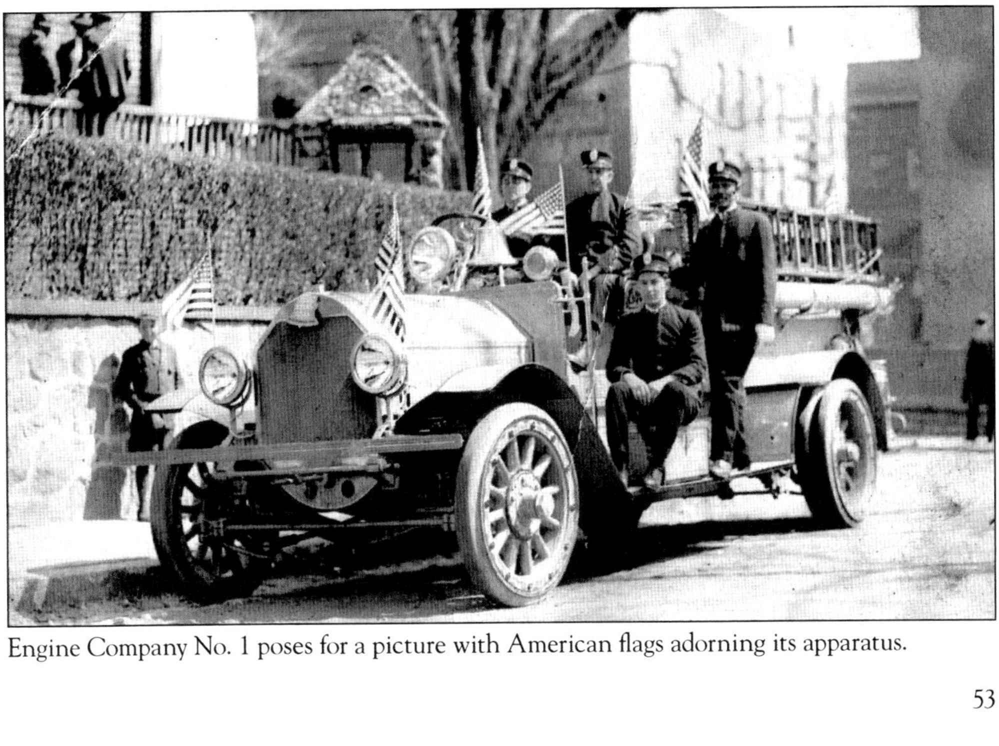

Engine Company No. 1 poses for a picture with American flags adorning its apparatus.

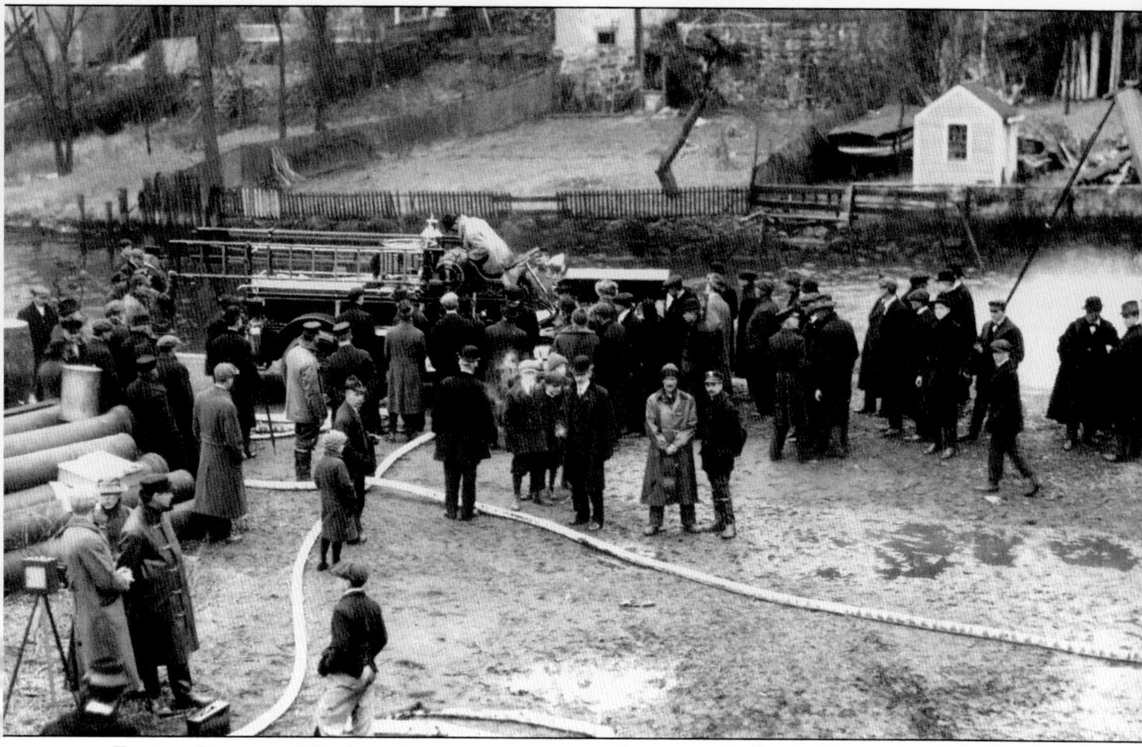

Engine Company No. 1 performs a capacity test for a crowd of onlookers around 1916.

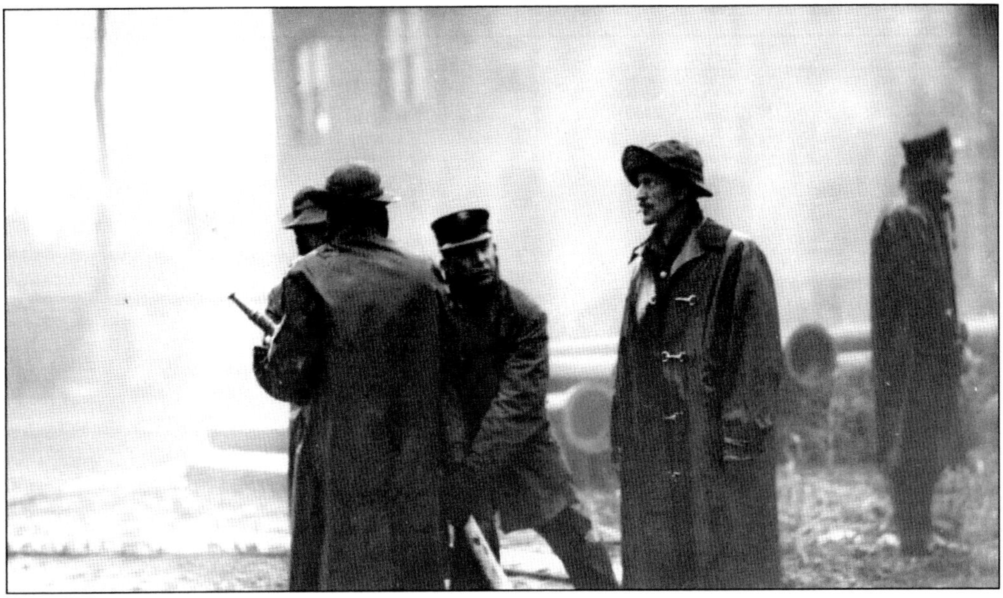

Firefighters operate multiple hand lines from the 1915 Seagrave engine. The fire apparatus is shown in the photograph below drafting from the Yantic River.

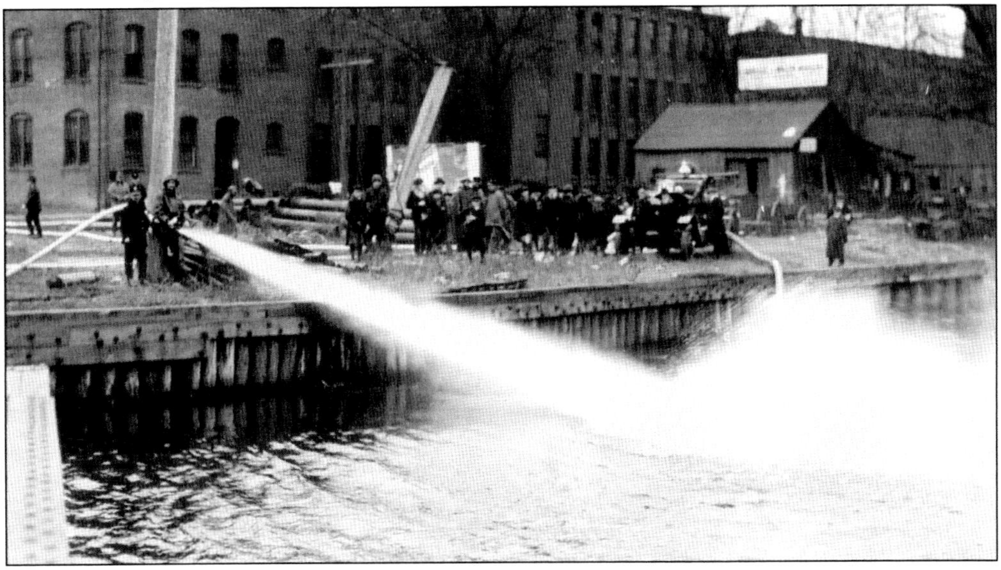

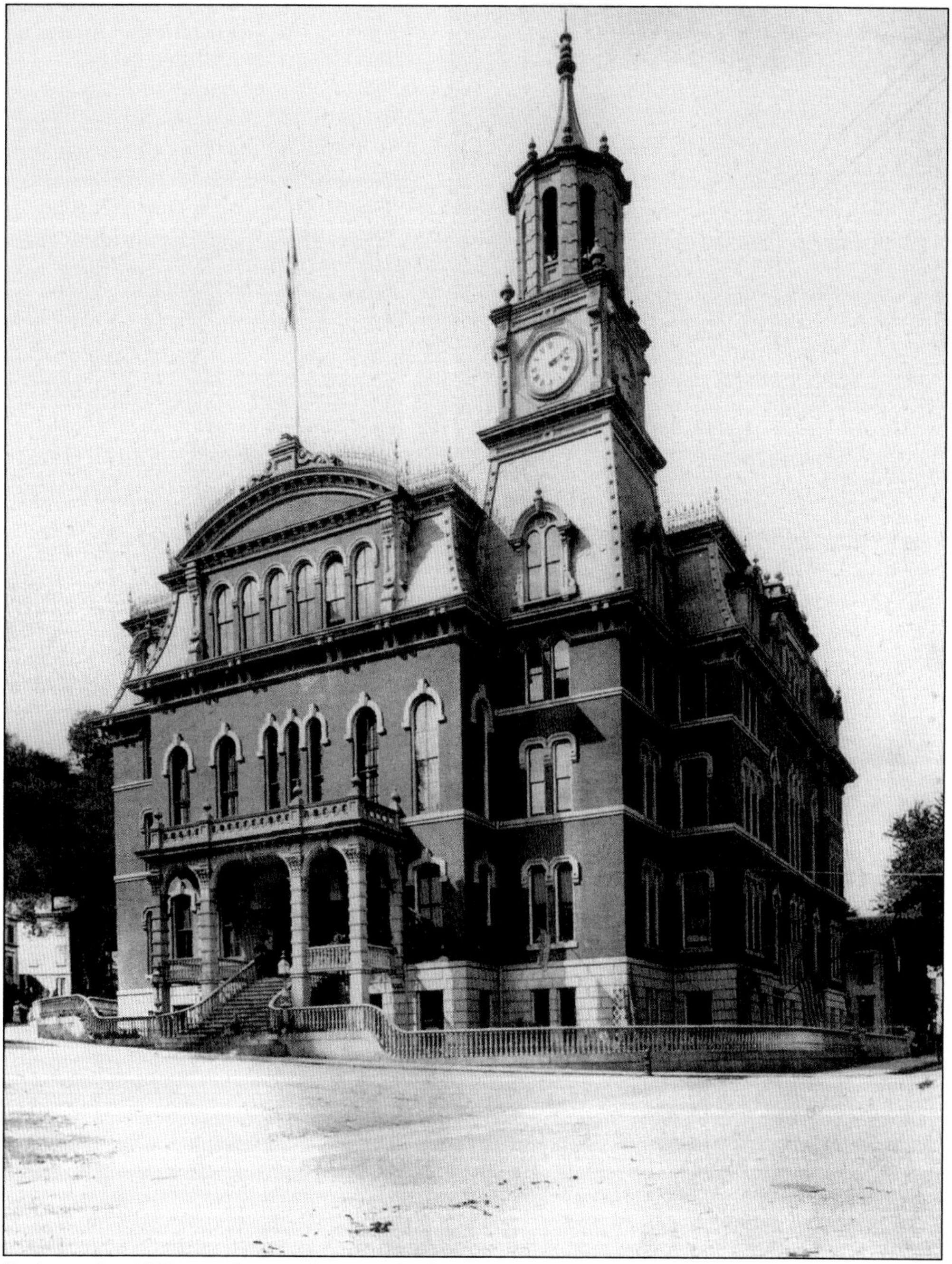

Built in the 1870s on the northern end of Union Square, city hall was the site of many fire demonstrations in the early 1900s. Fire department officials, including the chief and his assistants, utilized offices in city hall prior to the construction of headquarters on Chestnut Street.

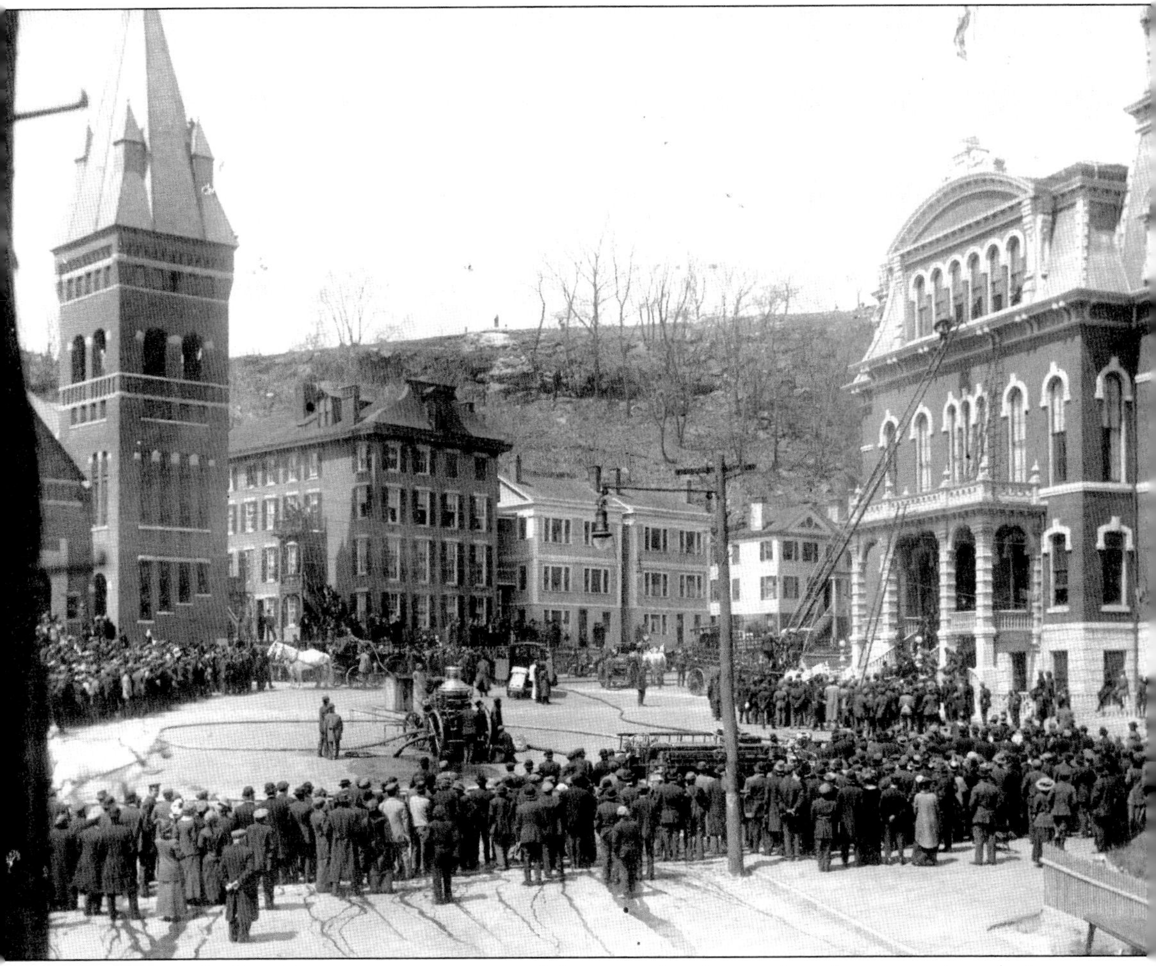

The Norwich Fire Department is pictured here performing a fire demonstration during Merchants Week in April 1914. The steam engine located in the center of Union Square is drafting from a large water vessel known as a cistern. These cisterns were located throughout the downtown area to supplement the hydrant system.

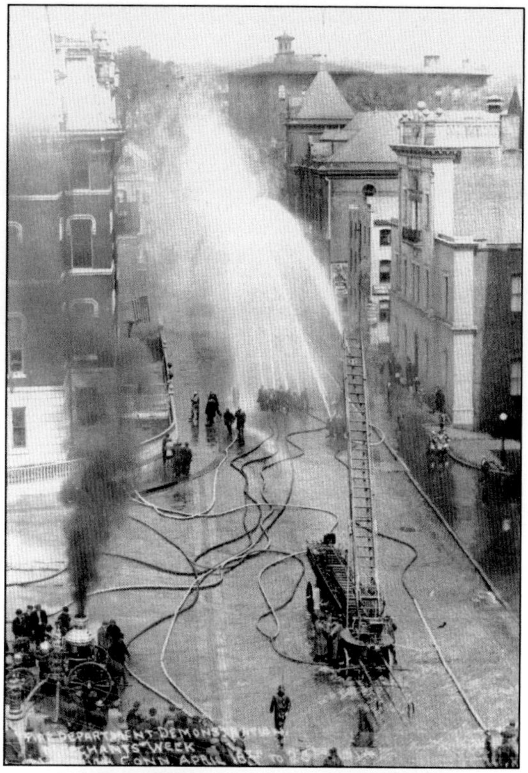

At left, multiple hand lines and master streams are pointed up Broadway in a demonstration of water flow during Merchants Week in April 1914. Below, a hand line is positioned in front of city hall to demonstrate the reach of fire streams on the taller buildings of the downtown area.

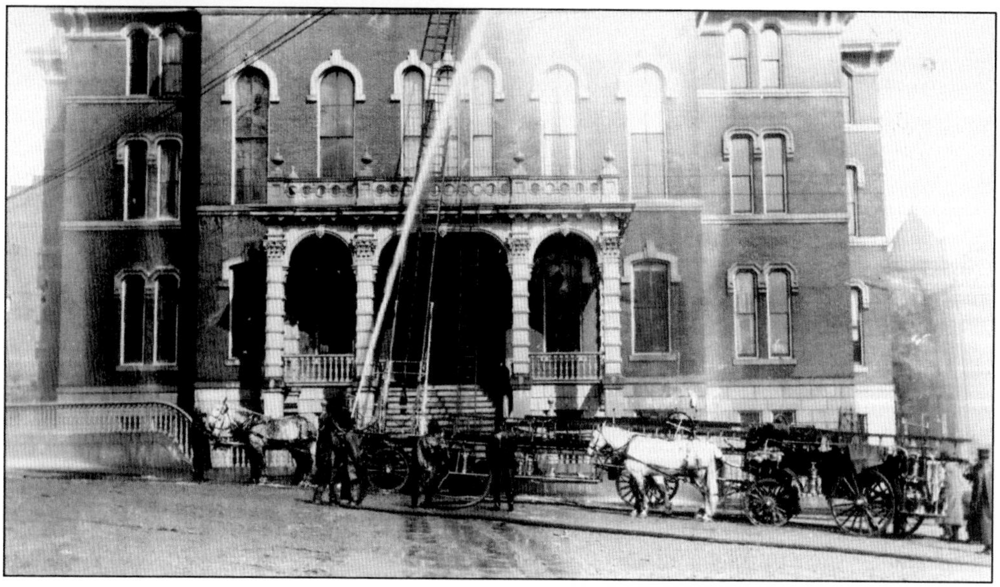

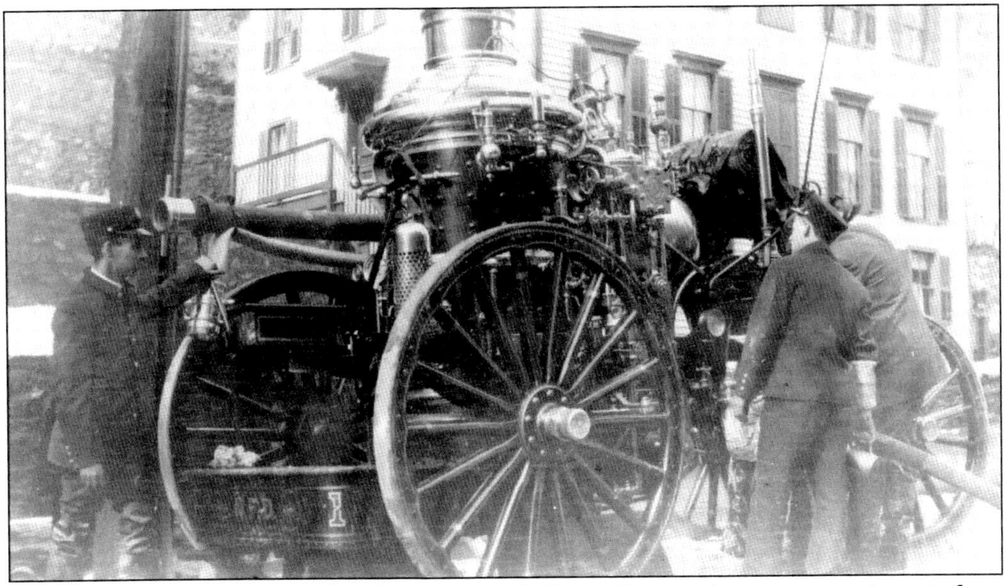

Fire department engineers tend to the American LaFrance Metropolitan steam engine as it flows water at a scene. This apparatus was designated engine No. 1 until 1915 when the department purchased a Seagrave tripple-combination chemical, pump, and hose car and named the new apparatus engine No. 1.

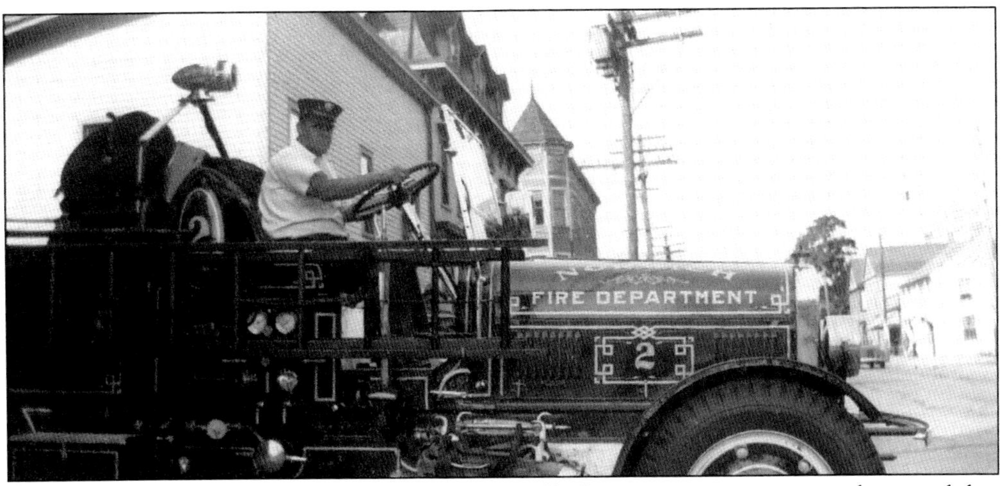

In 1922, the Norwich Fire Department converted all of its apparatus to motorized automobiles. This event marked the end of the horse-drawn era. In that same year, Engine Company No. 2 and Truck Company No. 2 were staffed with full-time firemen, bringing the Norwich Fire Department's apparatus numbers to six full-time engine companies and two full-time truck companies. Firefighter Joseph Barwell is pictured at the wheel of engine No. 2 in this 1940 photograph.

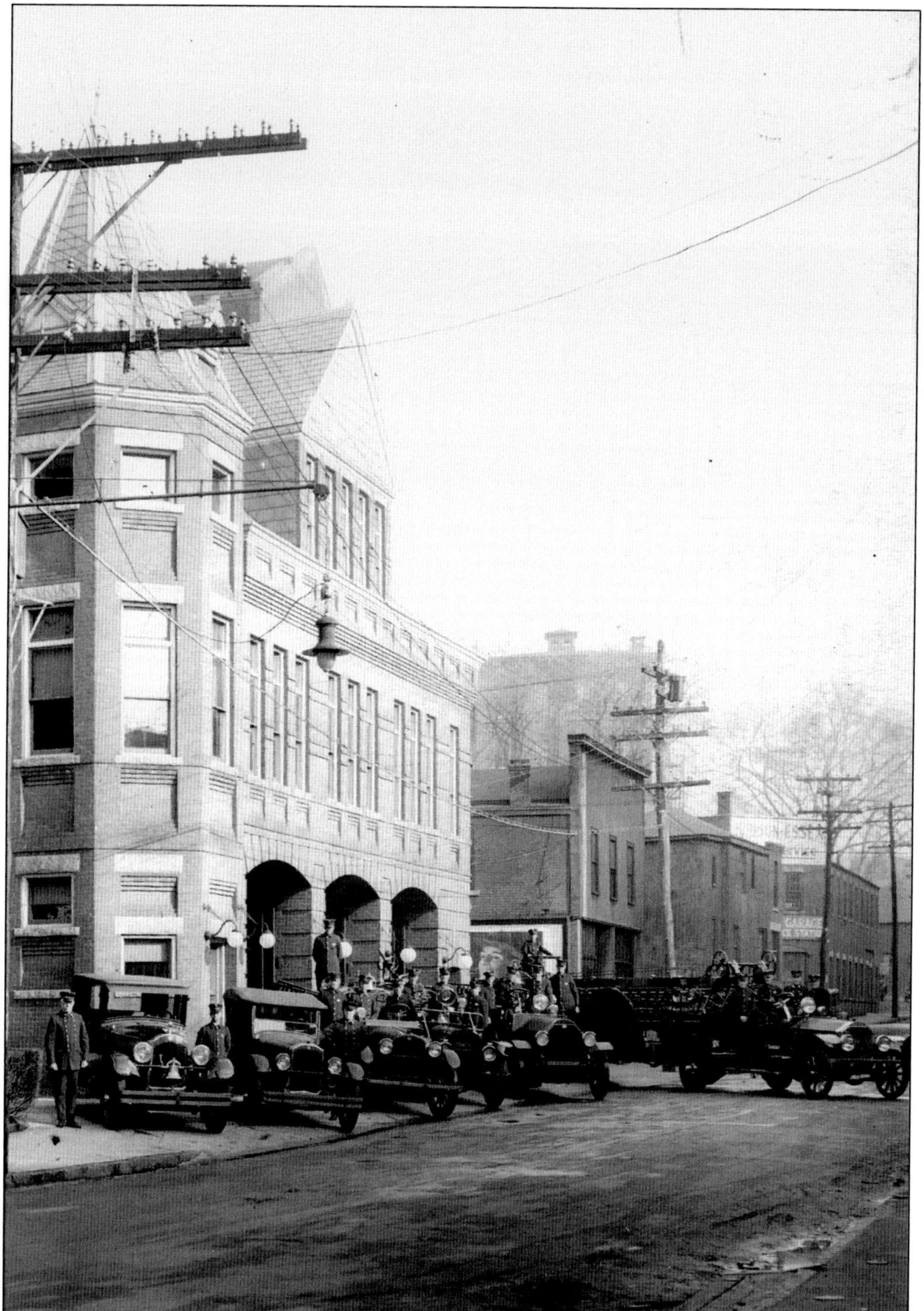

Fire companies from headquarters pose with their apparatus in this 1927 photograph. From left to right are Chief Howard L. Stanton, Stanton's chauffeur, and Deputy Chief Henry R. Taft (next to the two left automobiles).

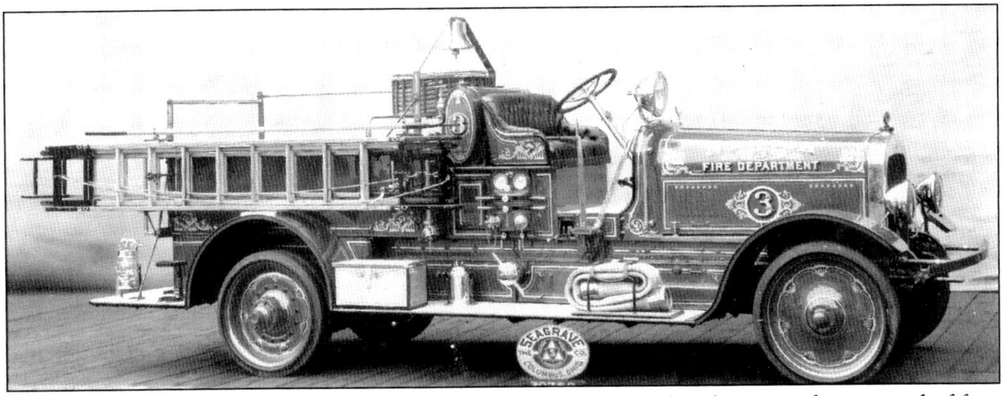

Engine Company No. 3 was formed in 1922 and operated from headquarters for a period of four years. In 1925, a new station No. 3 was built on Joseph Perkins Road adjacent to the Norwich Free Academy. The firehouse was designed to blend in with the surrounding homes.

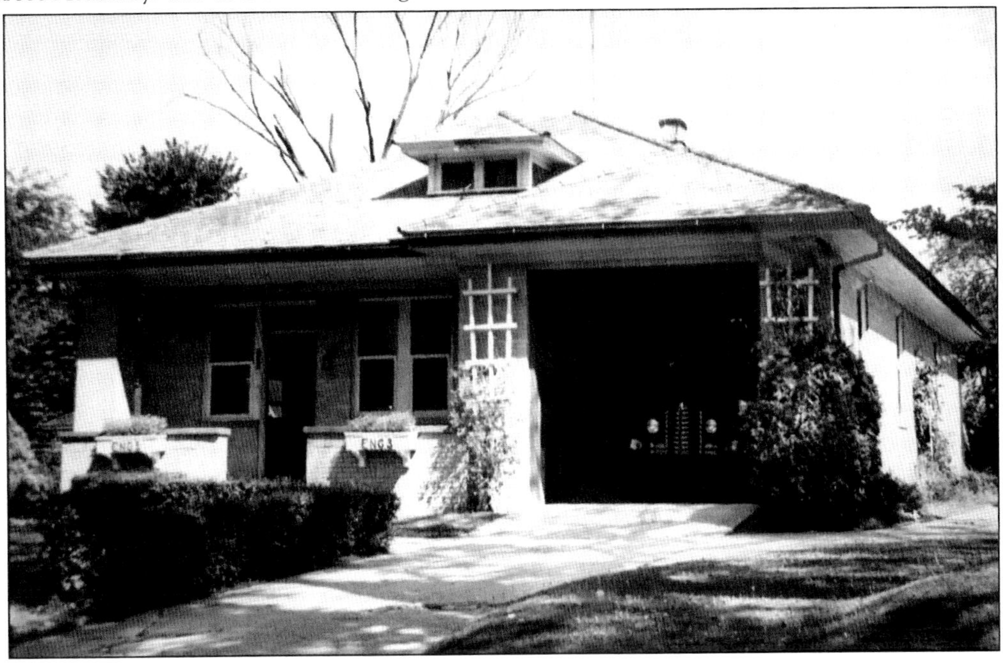

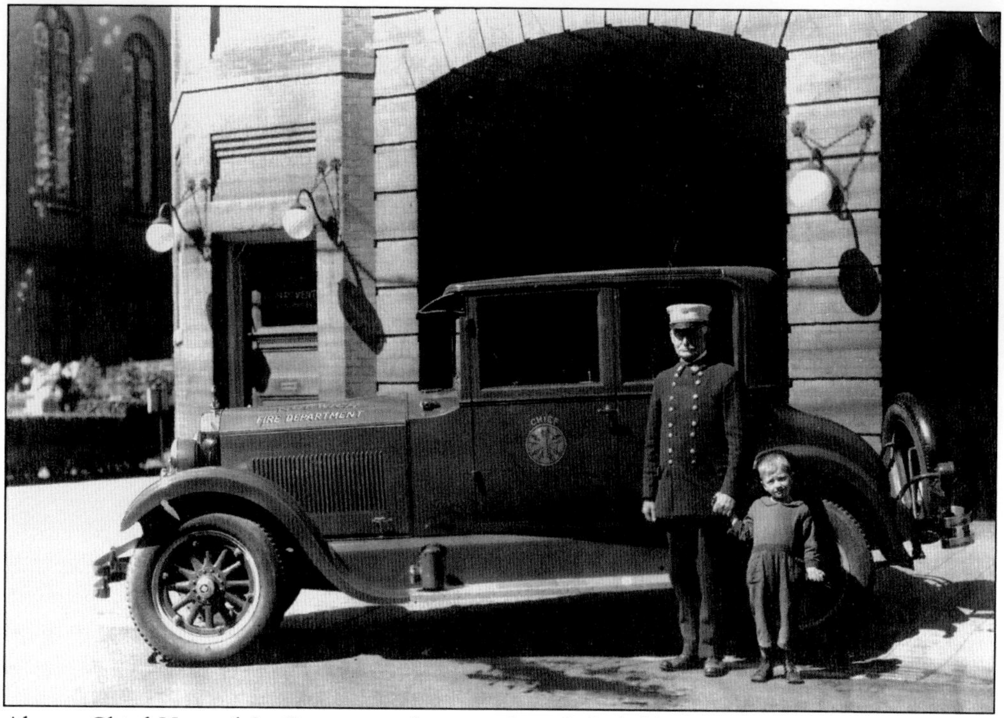

Above, Chief Howard L. Stanton and an unidentified child pose next to the chief's 1926 REO automobile. Below, the REO automobile is parked outside one of the apparatus bays of headquarters.

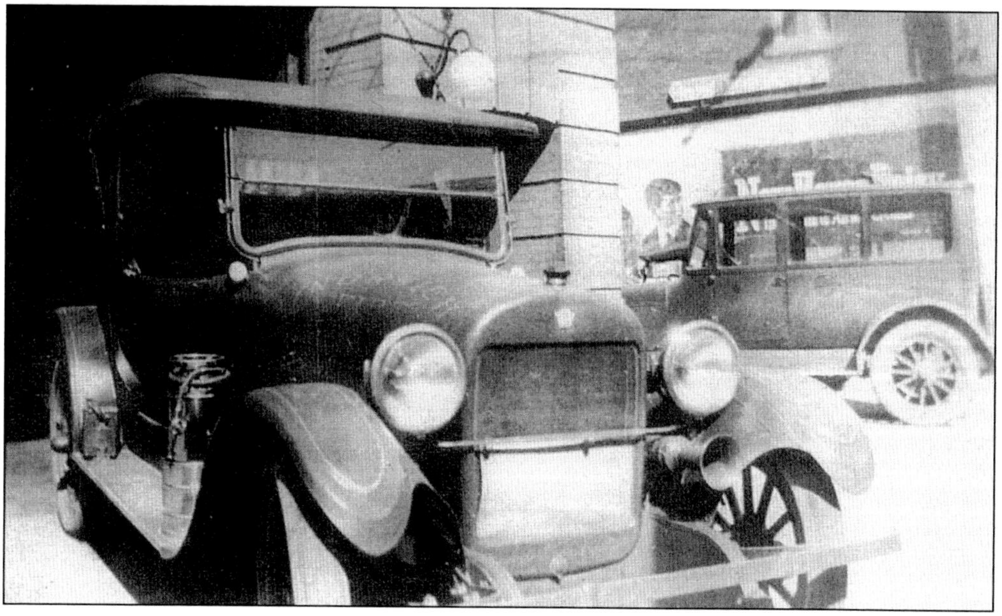

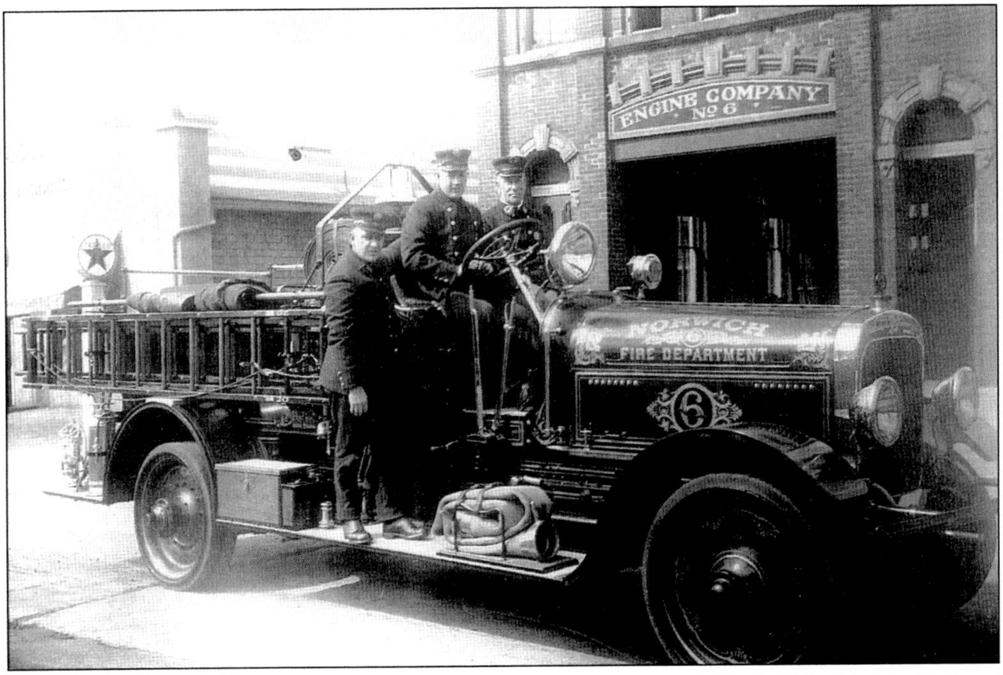

In 1922, Engine Company No. 6 was established in the old Thamesville firehouse, which was previously home to Independence Hose Company No. 6 in 1872. Above, Capt. Archibald Cochran, seated to the right of the photograph, poses with two unidentified Norwich firemen outside their station. Shown in its entirety at right, station No. 6 was built with a large and ornate bell tower that was removed in the Hurricane of 1938.

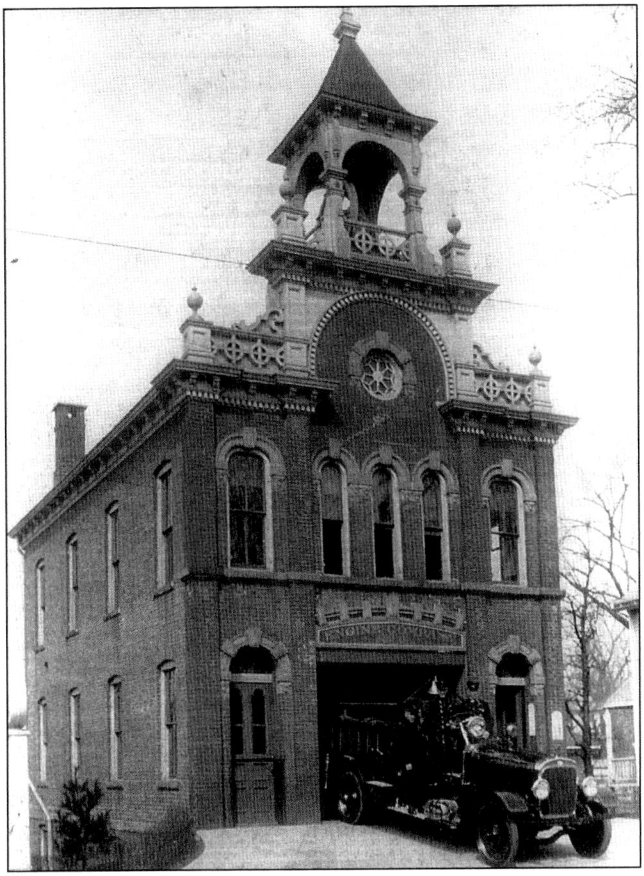

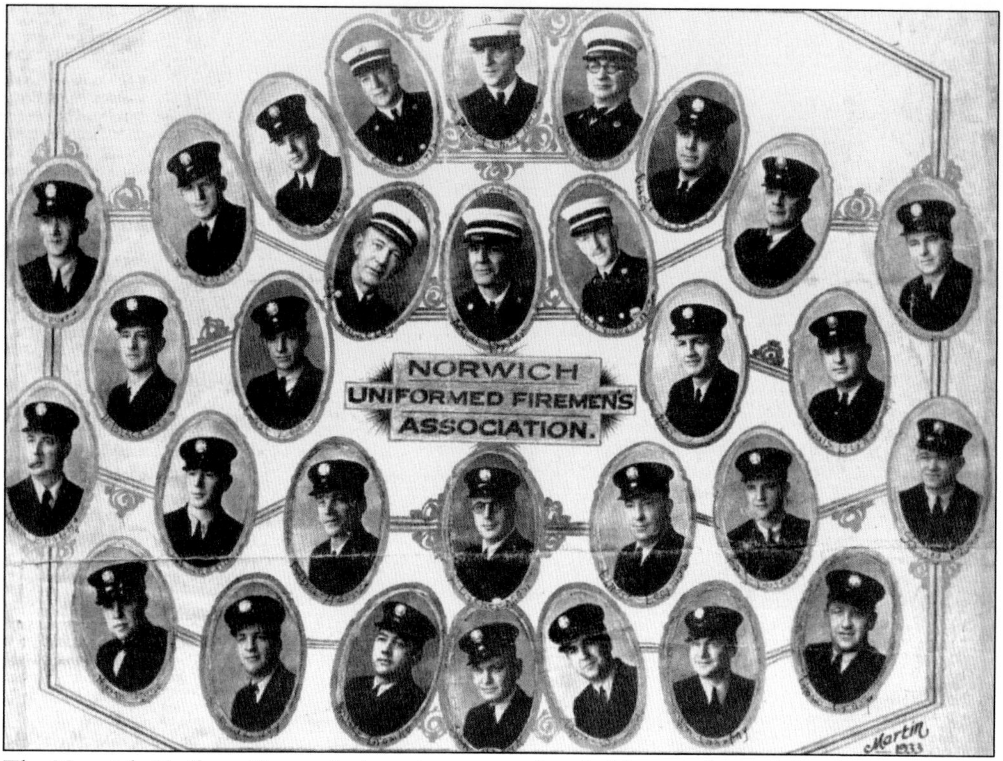

The Norwich Uniform Fireman's Association was founded by members of the department and existed for the purpose of negotiating equal and fair contacts with the City of Norwich. The association operated in much of the same manner as today's fire department union, Local 892. Pictured in the photograph above from left to right are the following: (first row) Marvin Hanna, Tom Kearney, Walter Gromko, John Ulrich, Thomas Carberry, William Confrey, and Tom Tracy; (second row) Adam Ensling, Charles Matthews, Albert Harris, Harold Leiberman, Elmore Leseman, Raymond Boisclair, and Stanley Vounce; (third row) Harry Blackburn, Forrest Lewis, (then title of association) Peter O'Neil, and Louis Blinderman; (fourth row) (above association title) Capt. Walter Huntley, Capt. Arthur Porter, and Capt. Robert Woodmansee; (fifth row) Dan Falvey, William Kearney, George Maloney, Capt. Archie Corcoran, Deputy Chief Dennis Brennan, Captain John Blackburn, George Barwell, John Nolan, and Everett Houston.

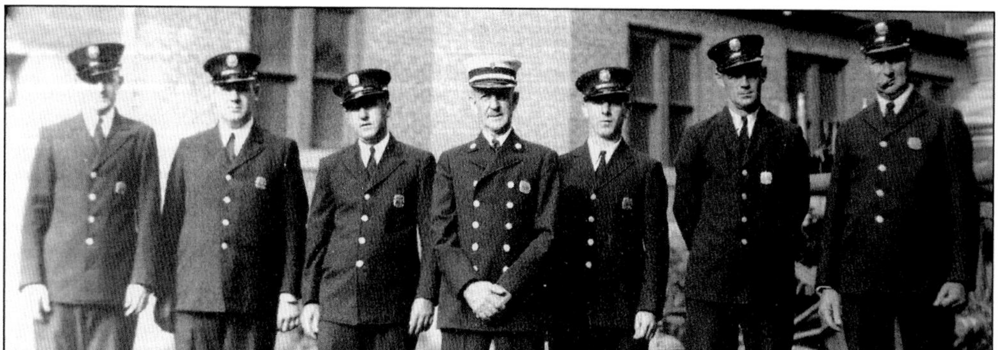

Pictured outside headquarters around 1934 are (from left to right) an unidentified fireman, George Maloney, Charles Matthews, Capt. Robert Woodmansee, Anthony Cosentino, Charles Tatro, and Adam Ensling.

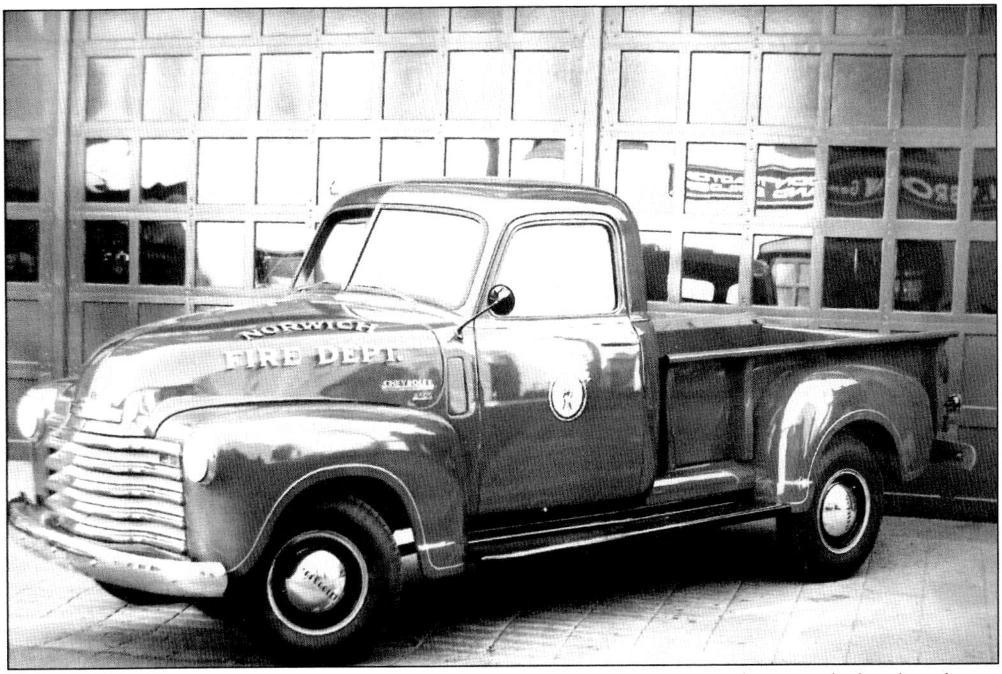

In the 1940s, the Norwich Fire Department received a Chevrolet pickup truck for the alarms division. The Norwich Fire Department serviced wires and pull stations of the fire alarm/telegraph system from the 1870s to the 1980s when the aging system was decommissioned.

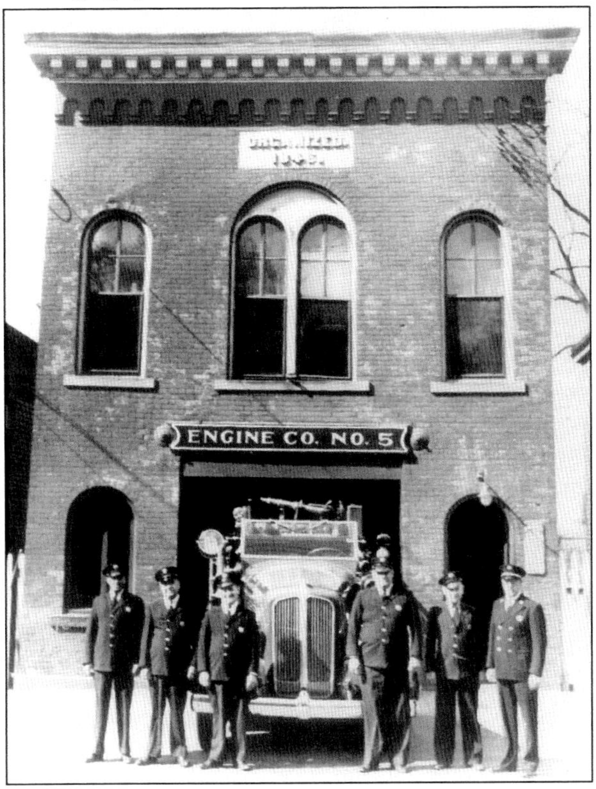

Engine Company No. 5 is pictured outside its station on West Main Street in April 1946. Included in the photograph are Capt. Harold J. Lieberman (far right), George Barwell, John Ulrich, Patrick Coleman, Thomas Carberry, and William LaRochelle.

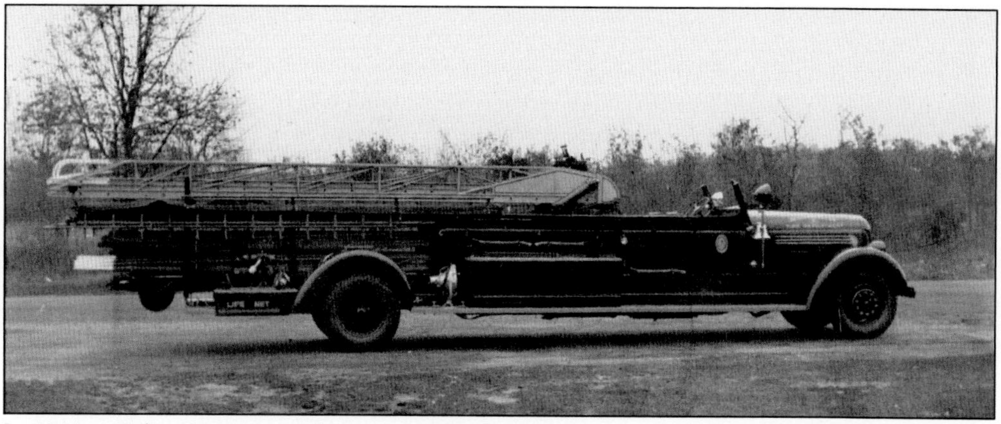

In 1944, a 65-foot Seagrave mid-mount ladder was purchased to replace Norwich's 1922 Seagrave tiller ladder. Below, an unidentified fireman gets the attention of an excited dalmatian puppy.

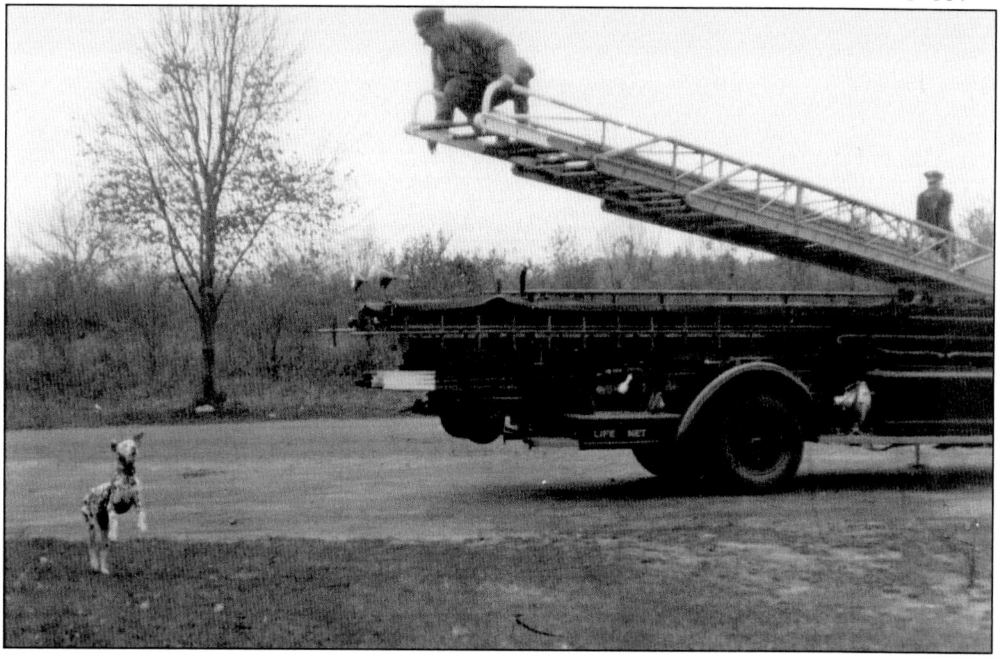

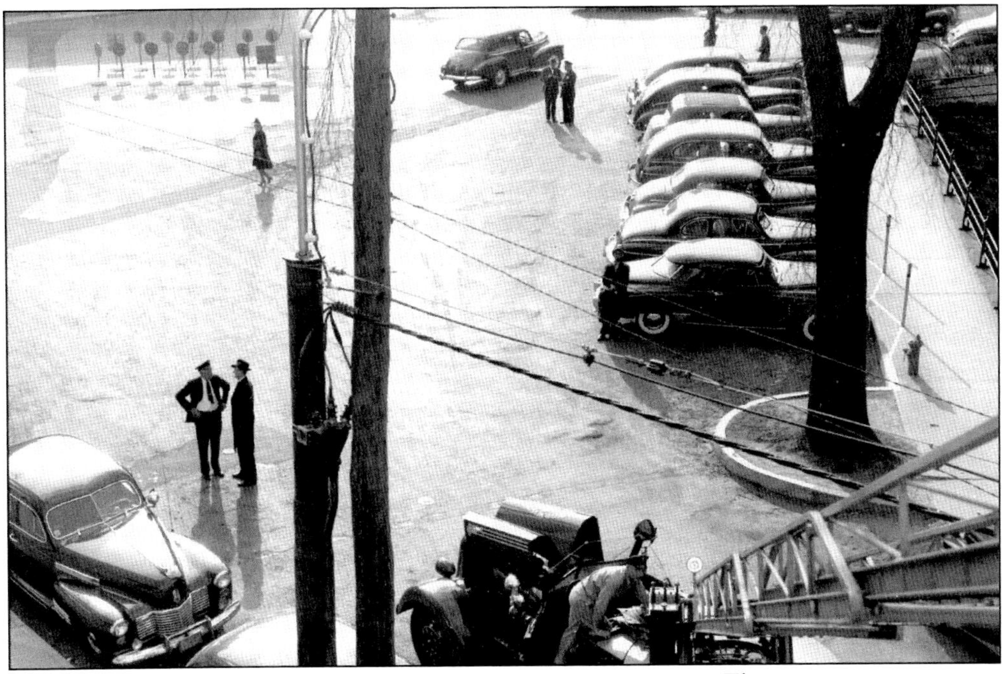

Aerial Truck Company No. 1 is shown operating in Union Square. This unique vantage point allows the layperson to see what it might be like to be a ladder man.

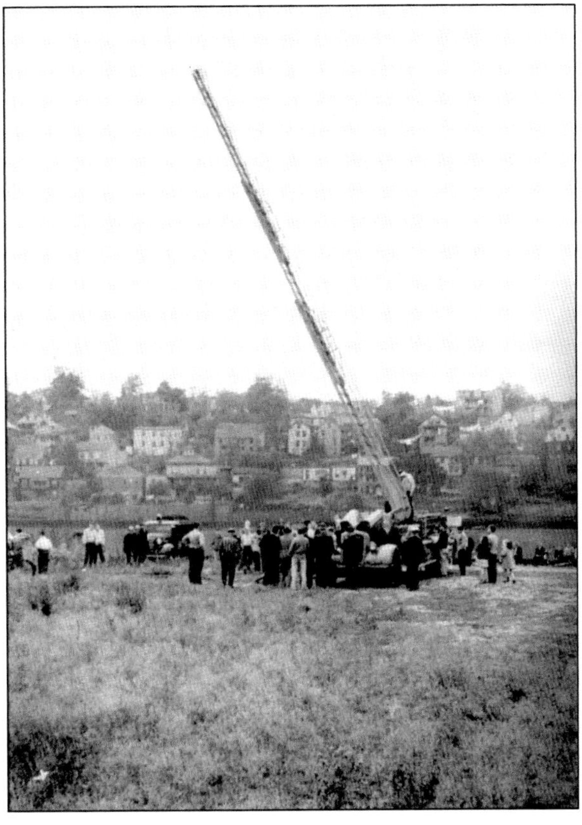

Aerial Truck Company No. 1, along with other Norwich units, is seen operating along the banks of the Thames River on September 9, 1944. The 65-foot Seagrave ladder is shown at full extension.

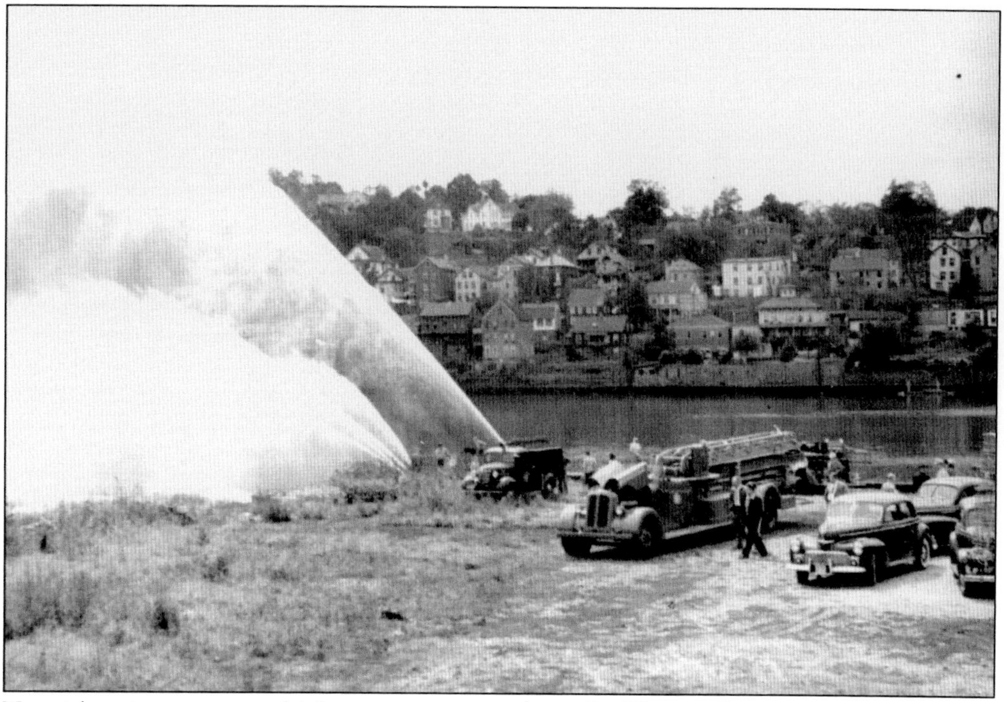

Norwich units operate multiple master streams along the Thames River in 1944. Below, the Norwich Fire Department's emergency defense pumper makes its debut. It is uncertain if this pumper was intended for firefighting, dewatering, or a potential large-scale attack by enemy forces during World War II.

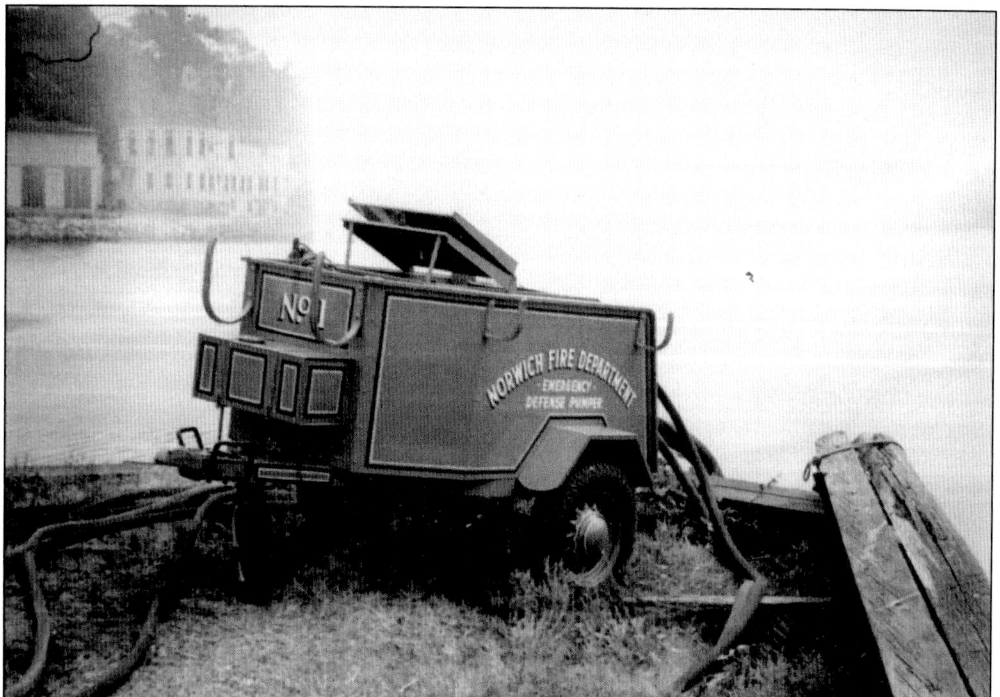

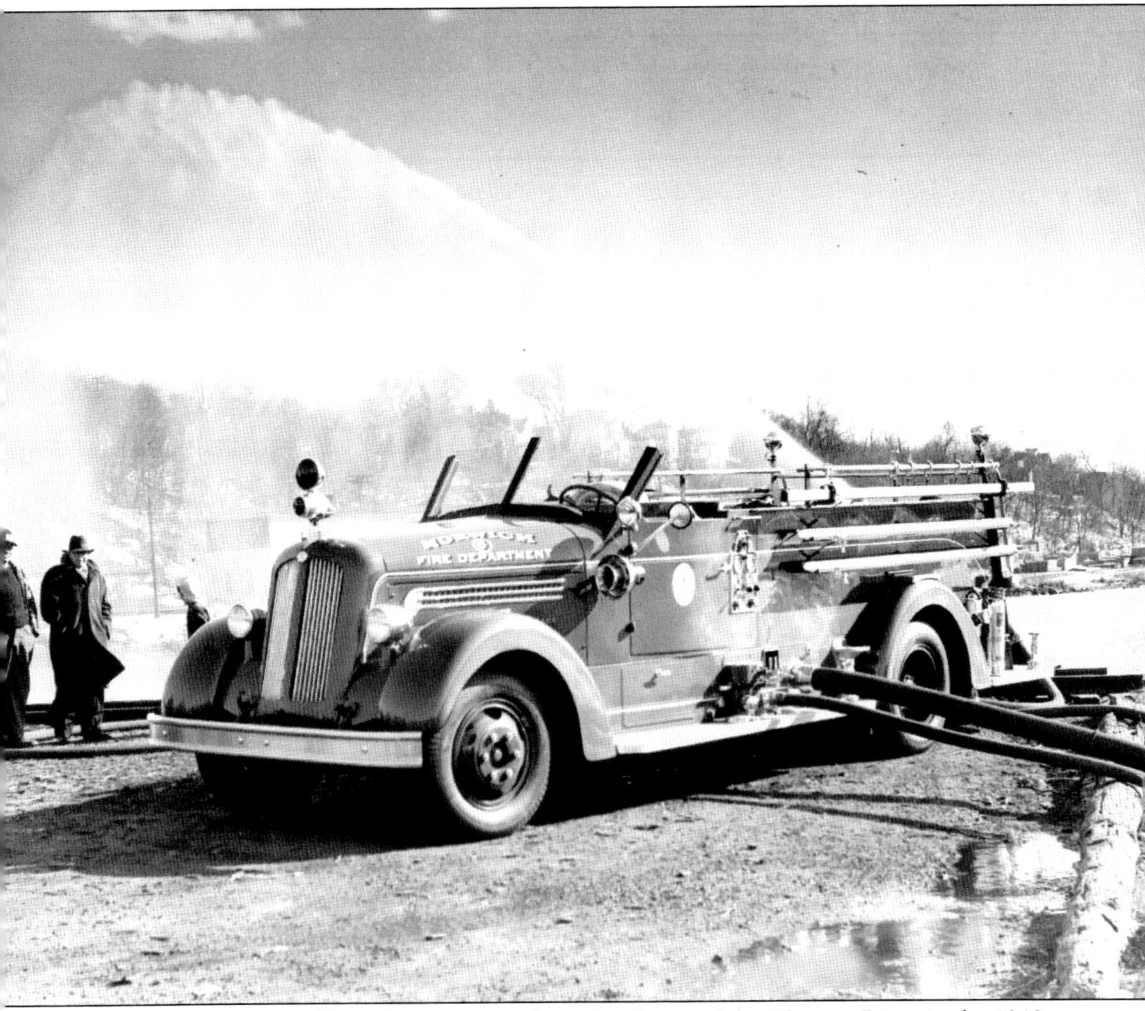

Engine No. 6 is pictured here flowing water from the shores of the Thames River in the 1940s. This early-1940s model Seagrave engine was stationed in the Thamesville District of Norwich.

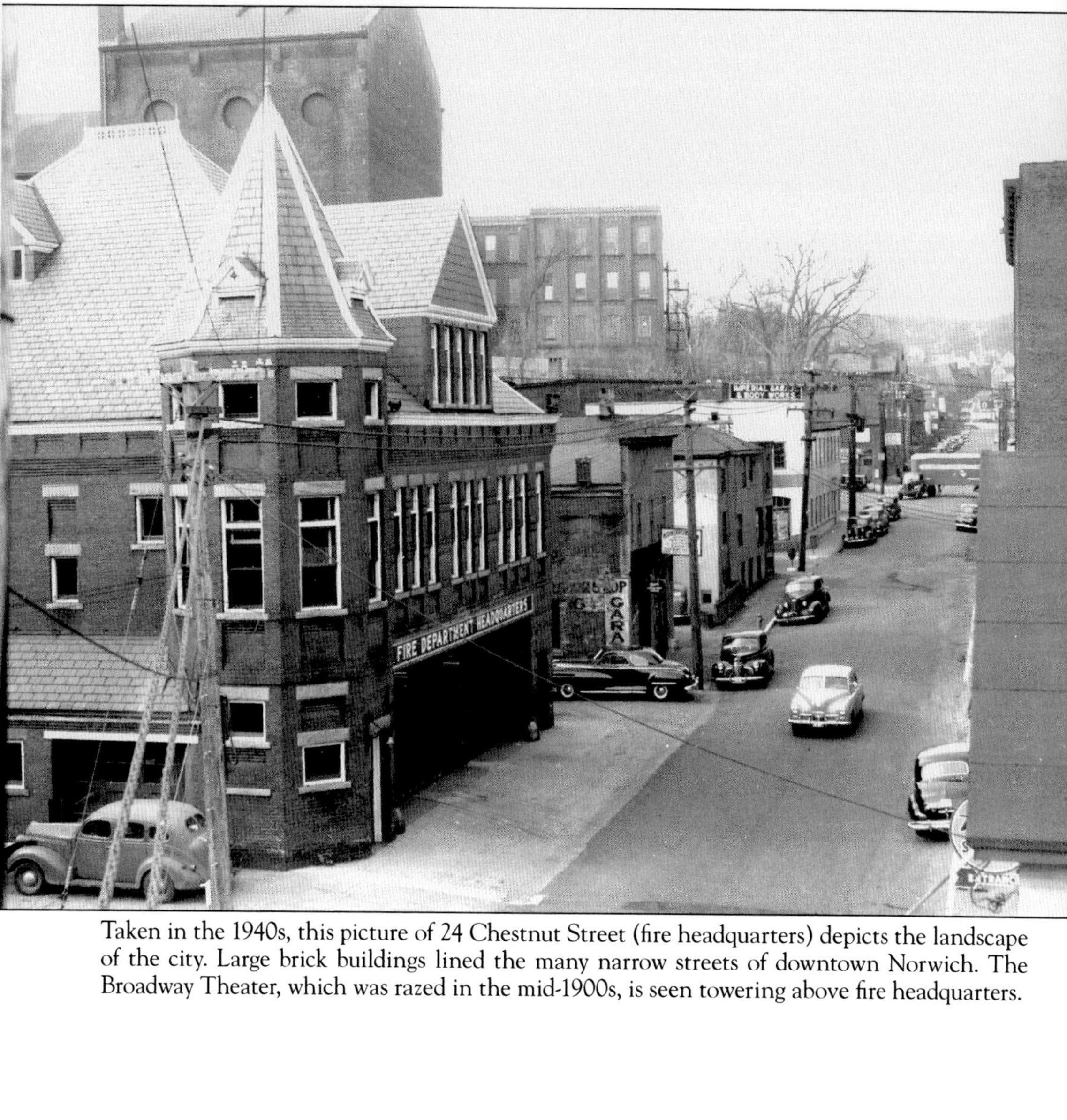

Taken in the 1940s, this picture of 24 Chestnut Street (fire headquarters) depicts the landscape of the city. Large brick buildings lined the many narrow streets of downtown Norwich. The Broadway Theater, which was razed in the mid-1900s, is seen towering above fire headquarters.

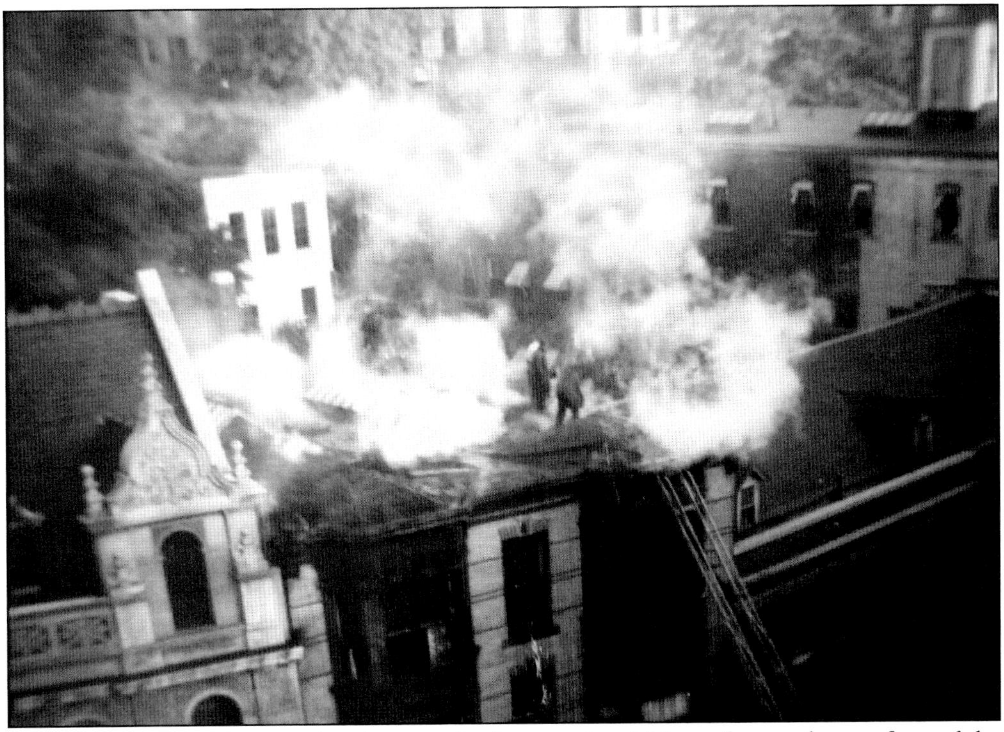

Firefighters are seen here in the photograph above extinguishing a fire on the top floor of the Martin House, located at 16–22 Broadway. Below, aerial truck No. 1 provides roof access from the middle of Broadway.

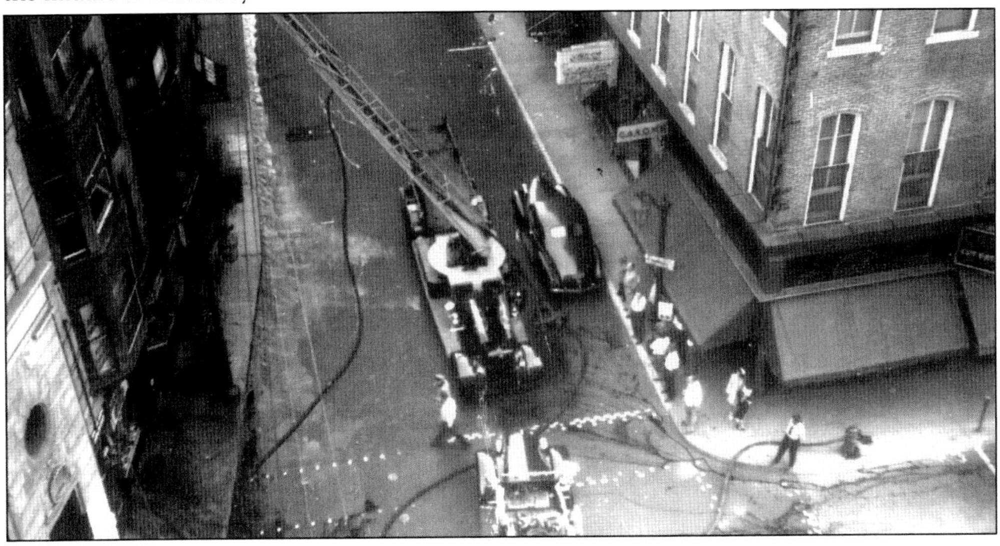

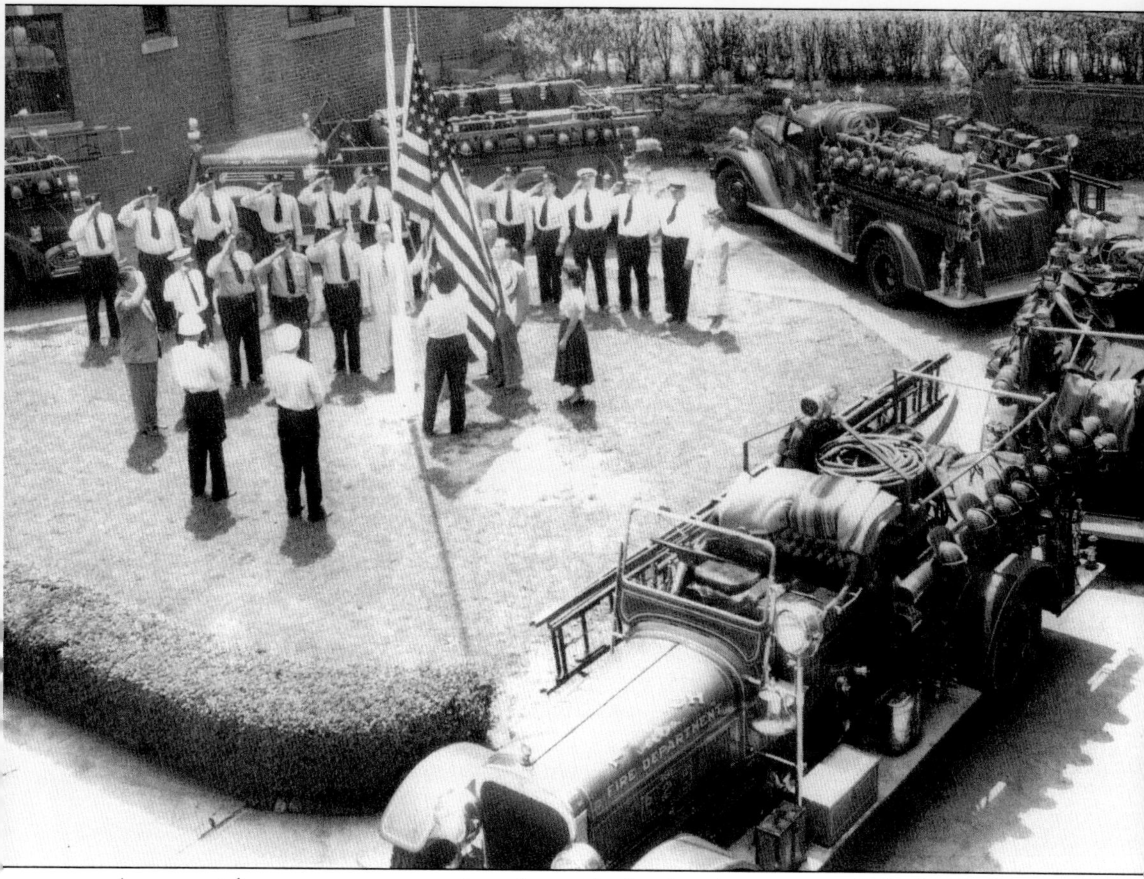

A memorial ceremony was conducted outside fire headquarters in 1953 for firefighter Bill Lafleur's two sons who were lost in the Korean War. Engine No. 2 is pictured in the foreground of this photograph.

Three

THE MODERN ERA

The latter half of the 20th century saw many improvements to the quality of fire protection and emergency care. From the formation of a rescue company and an emergency medical service to fire prevention education and today's special hazards, the Norwich Fire Department has been able to adapt to meet the needs of the public on every level.

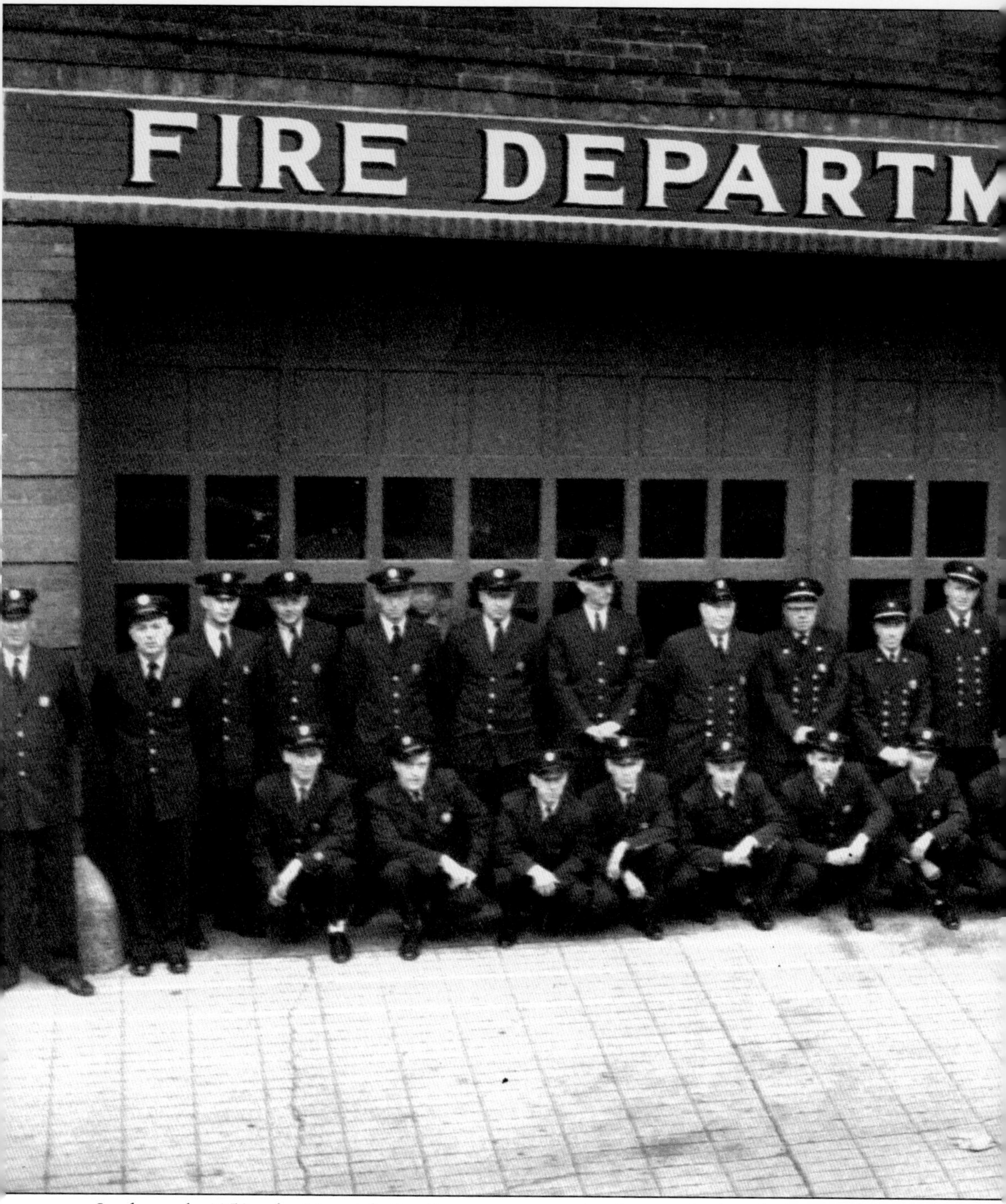

In the early 1950s, the Norwich Fire Department received a new look with the purchase of new uniforms. The men of the Norwich Fire Department pose in front of fire headquarters, located at 24 Chestnut Street. Pictured from left to right are (first row) A. Consentino, L. Counihan, W. Brewer, W. Wilson, J. Colman, R. Herz, C. Murkett, A. Ellison, S. Piechowski, E. Romano, W. Lafleur, C. Cobb, T. Carberry, E. Moran, J. Boisclair, H. Lamphere Jr., J. Humphrey, A. Mageski, W. Dougherty, and A. Connell; (second row) G. Driscoll, H. Milewski,

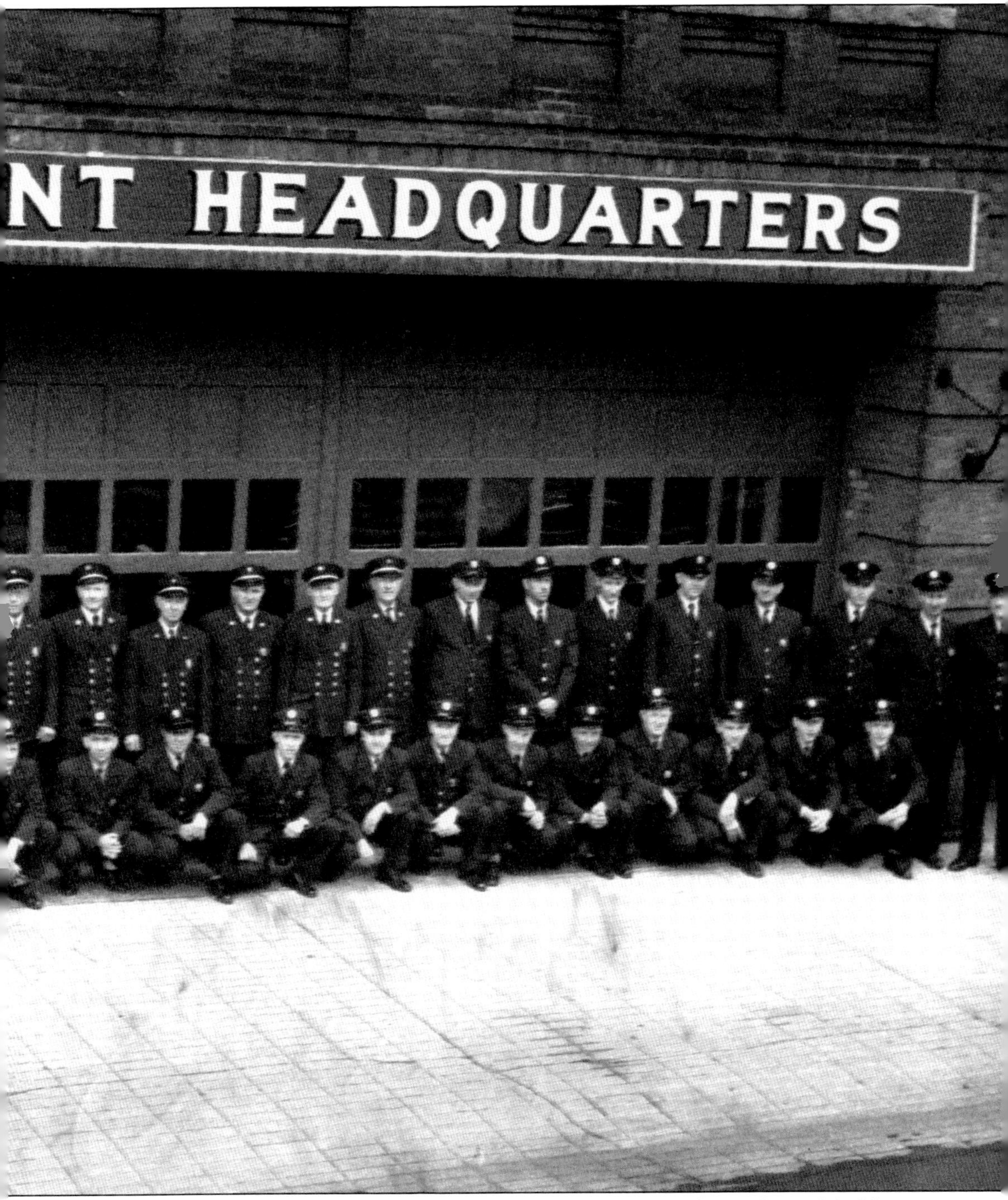

T. LaFrenier, E. Zulka, J. Burke, H. Coburn, S. Osik, R. Wilson, Capt. M. Hanna, Capt. E. Leseman, Capt. W. Burns, Capt. P. Coleman, Deputy W. LaRochelle, Chief W. Confrey, Deputy C. Matthews, Captain A. Chmielewski, Captain W. Kearney, Capt. H. Ludlow, E. Jennes, W. Bohara, G. Faulkner, C. Smith, T. Twomey, P. Bialowas, W. Sheridan, and E. Breen. The men inside the building are Cal Kirshner and John Ulrich.

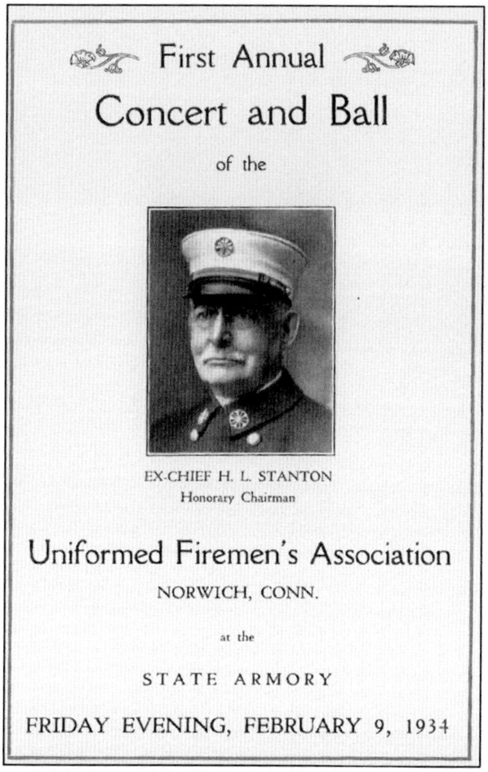

First Annual

Concert and Ball

of the

EX-CHIEF H. L. STANTON
Honorary Chairman

Uniformed Firemen's Association

NORWICH, CONN.

at the

STATE ARMORY

FRIDAY EVENING, FEBRUARY 9, 1934

On February 9, 1934, the first annual Uniformed Fireman's Association Concert and Ball was held at the old Norwich Armory on McKinley Avenue. Department members, escorted by their significant others, attended this formal dance, which involved such styles as the fox trot and the waltz. Retired chief Howard L. Stanton and retired captain Alvin Lewis were guests of honor of this inaugural gala.

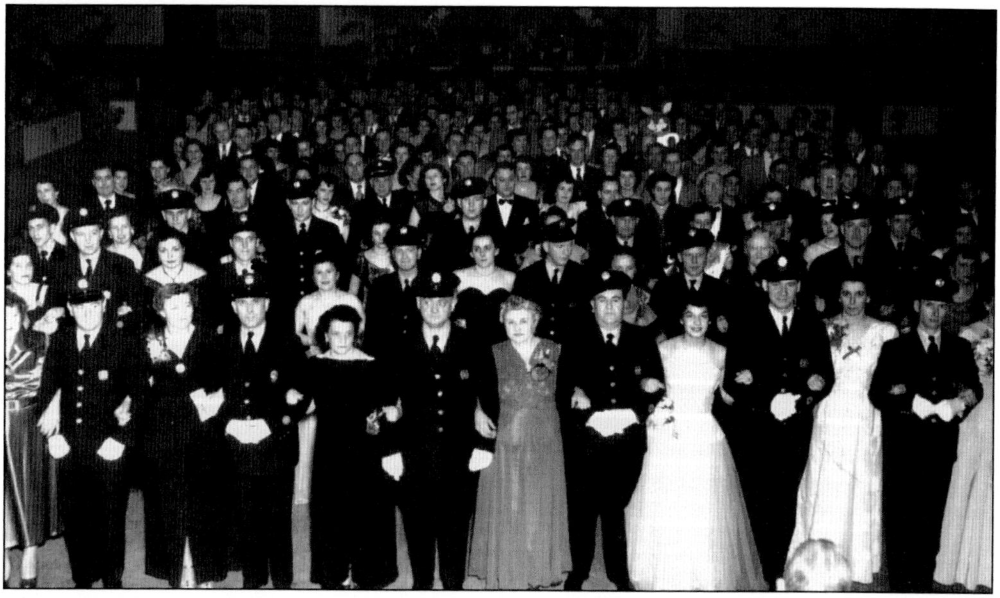

The annual firemen's ball was continued throughout the 1940s and 1950s. Firemen pictured here around 1953 are, from left to right, (first row) Charles Matthews, William LaRochelle, Louis Blinderman, Thomas Carberry, Anthony Cosantino, and Clarance Murkette; (second row) Cal Kirshner, Thomas LaRochelle, Edward Murrane, William Dougherty, Zigmund Piechowski, and Edward J. Breen; (third row) Cal Cobb, Peter Bialowas, Jerry Falkner, Charles Smith, Raymond Boisclair, John Ulrich, and Harold Lamphere Jr.; (fourth row) Chief Joseph Tracy.

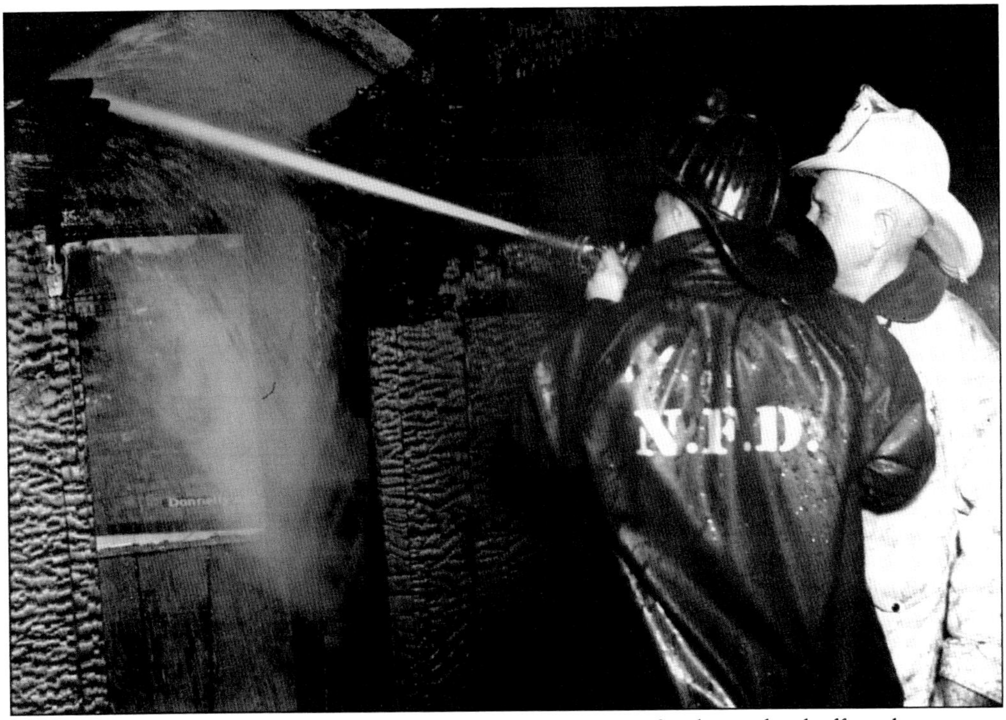

Chief William Confrey (right) and an unidentified fireman finish overhaul efforts by wetting down charred framing members.

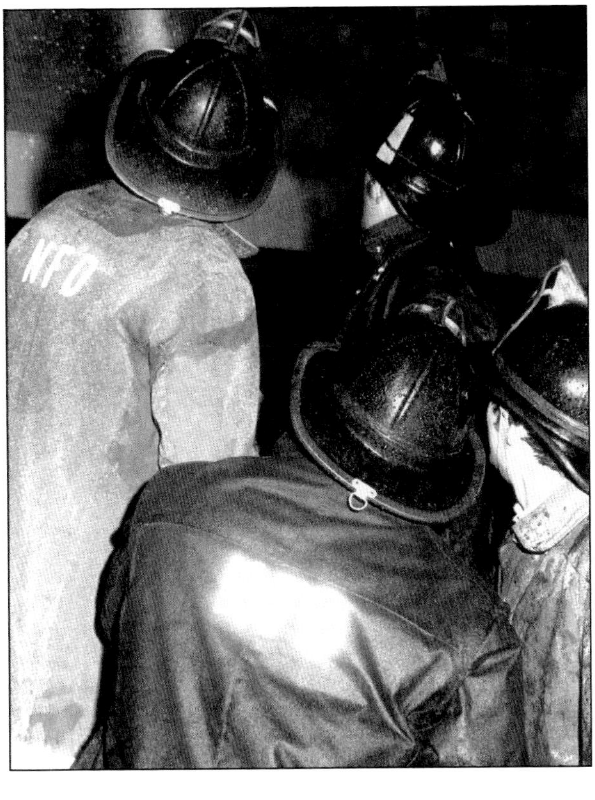

Four unidentified Norwich firemen help direct a two-and-one-half-inch hose line on the fire. This photograph typifies the firefighter persona of an untiring and anonymous guardian.

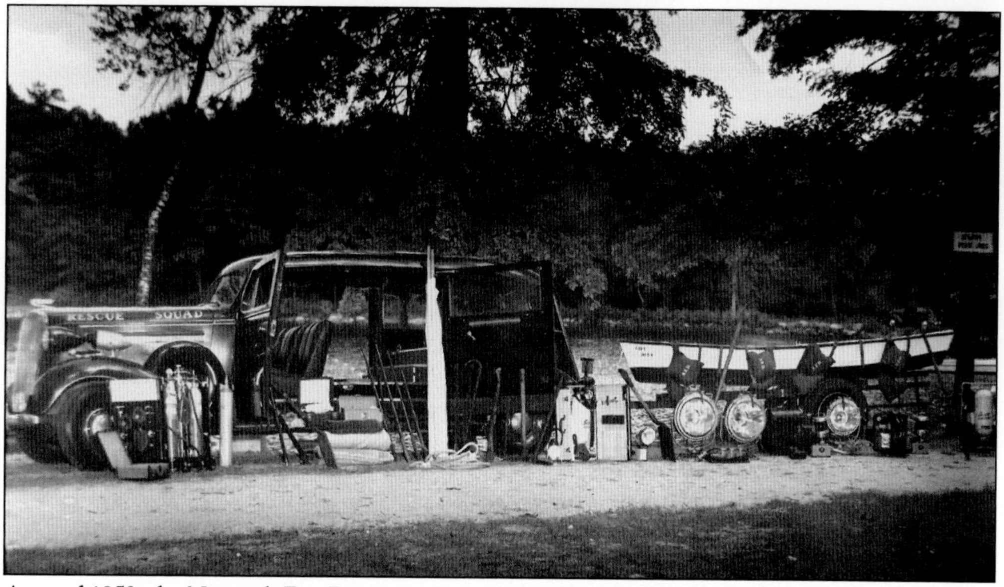

Around 1952, the Norwich Fire Department converted a 1930s Buick hearse into a rescue vehicle capable of patient transport. The apparatus was built by the members of the department and carried lifesaving medical equipment as well as specialty items used in the fire service.

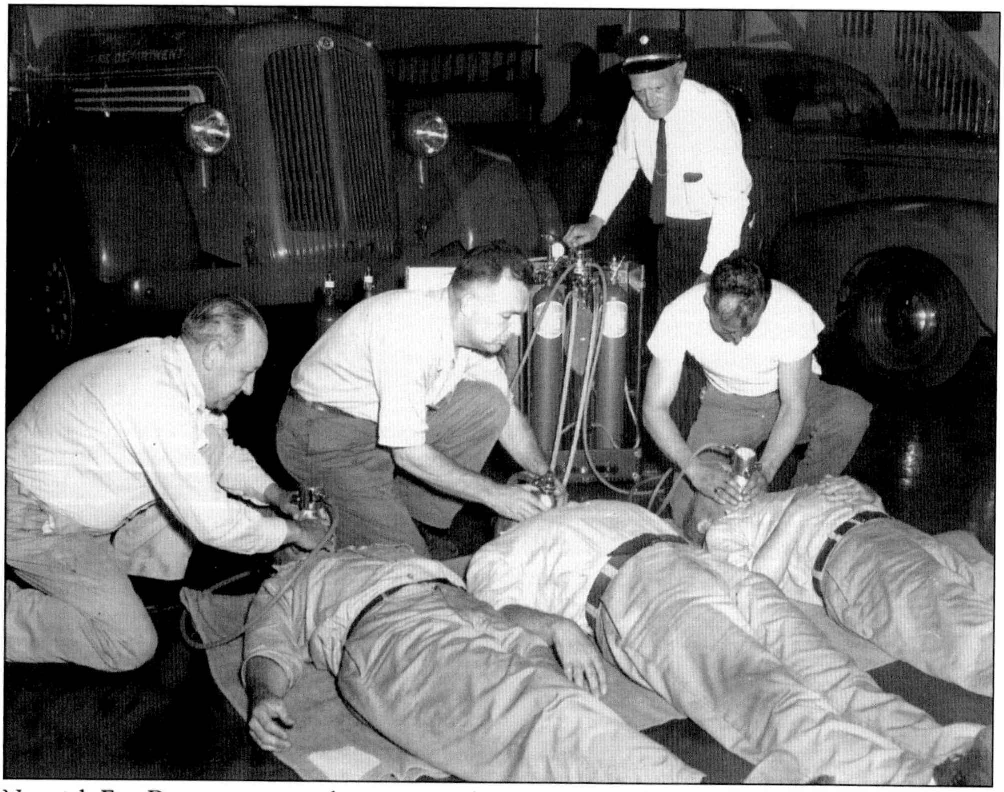

Norwich Fire Department members are seen here reviewing the use of pressure-assisted oxygen ventilators. From left to right are Capt. William Confrey, firefighter Henry Milewski, Deputy Chief Arthur Porter, and firefighter Thomas LaRochelle.

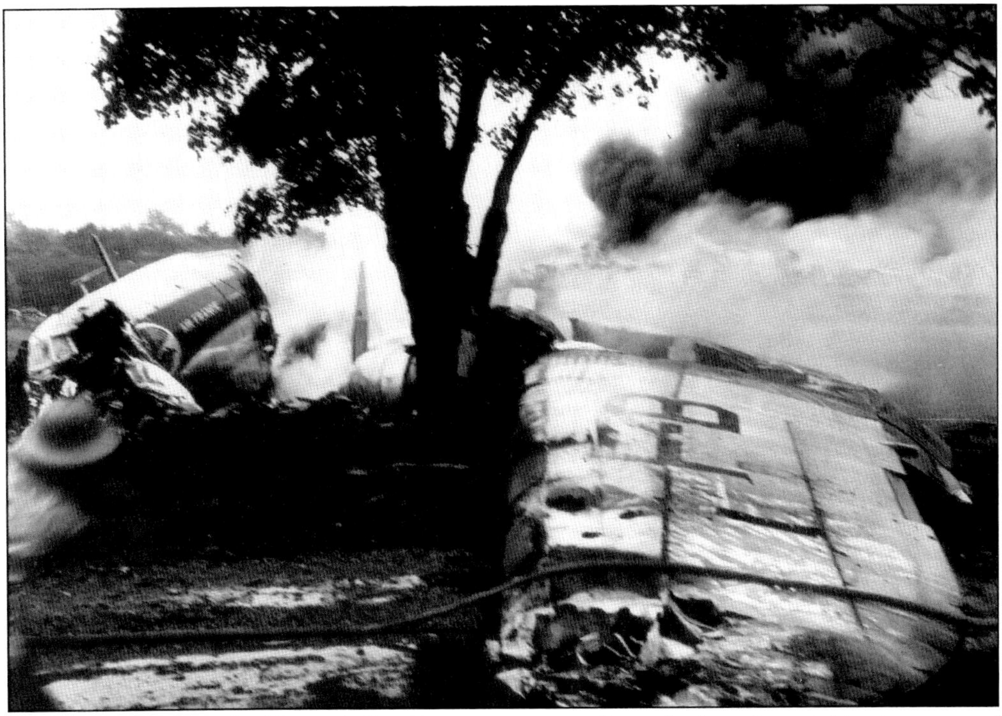

On August 4, 1954, the Norwich Fire Department was dispatched to the neighboring town of Preston to combat a large airplane fire. Surprisingly, all 37 passengers aboard the Air France Super Constellation airplane survived the crash.

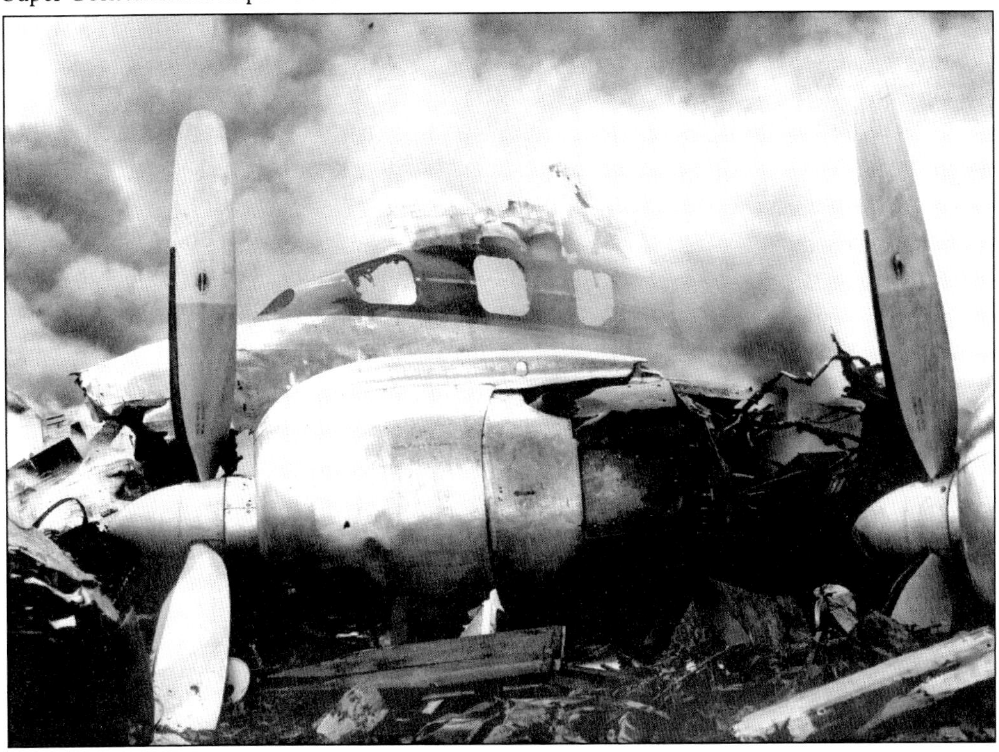

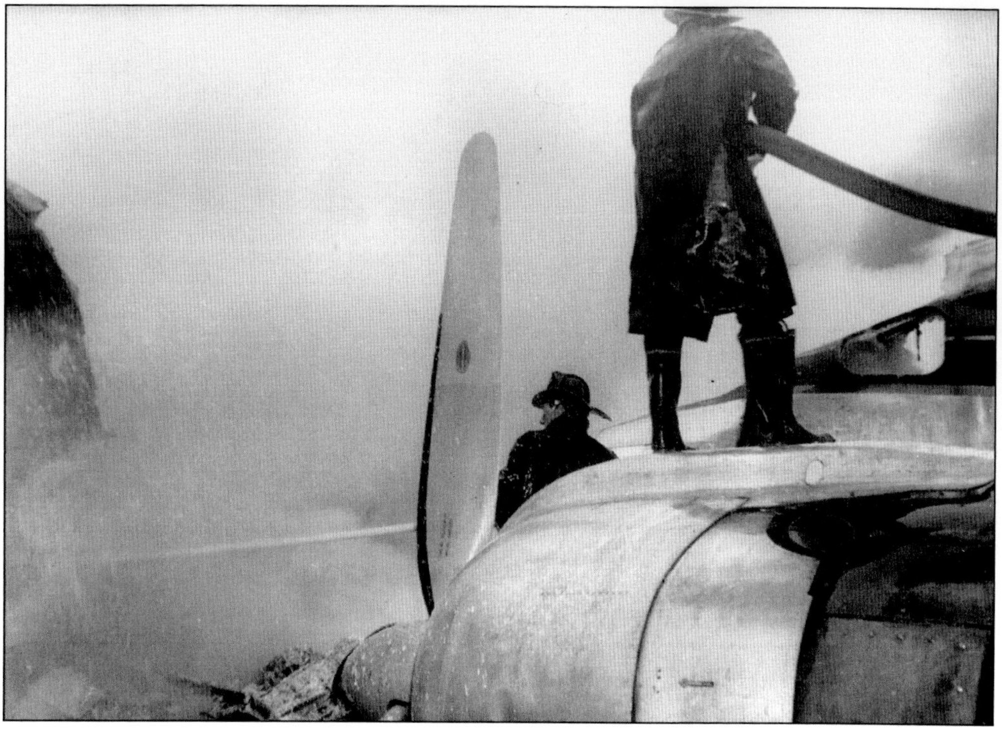

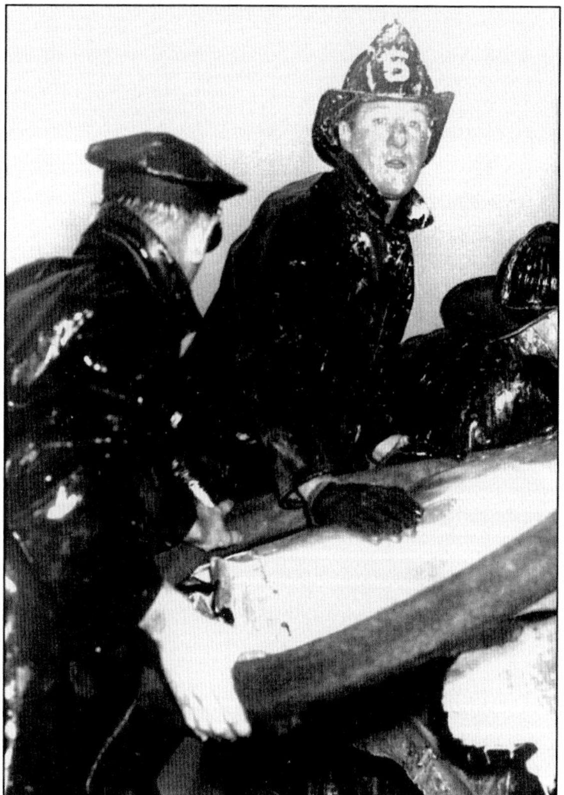

Above, firefighters are seen applying foam to the fuel-soaked fuselage of the Air France airplane. Eyewitness accounts of the fire indicate the aircraft engines contained a fare amount of magnesium, which proved to be difficult to extinguish. At left, firefighter Felix Lebrie is captured in action.

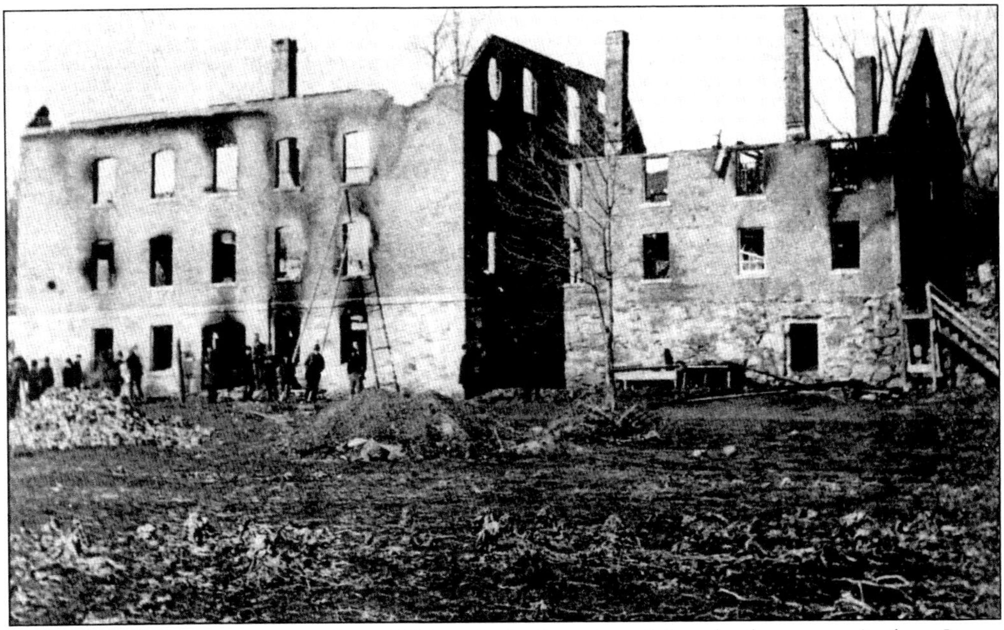

Norwich suffered a great tragedy in 1876 with the burning of the almshouse on Asylum Street (above). This three-story brick building was home to 55 inmates of the alms home, a refuge for the blind, handicap, and the mentally insane. The fire, which occurred on the evening of March 12, 1876, quickly spread from a heater on the first floor to the entire structure. The blaze claimed the lives of 14 individuals. Below, the almshouse was eventually rebuilt on the same premises. Having been vacant for a short time, the home was again destroyed by fire in the early 1950s.

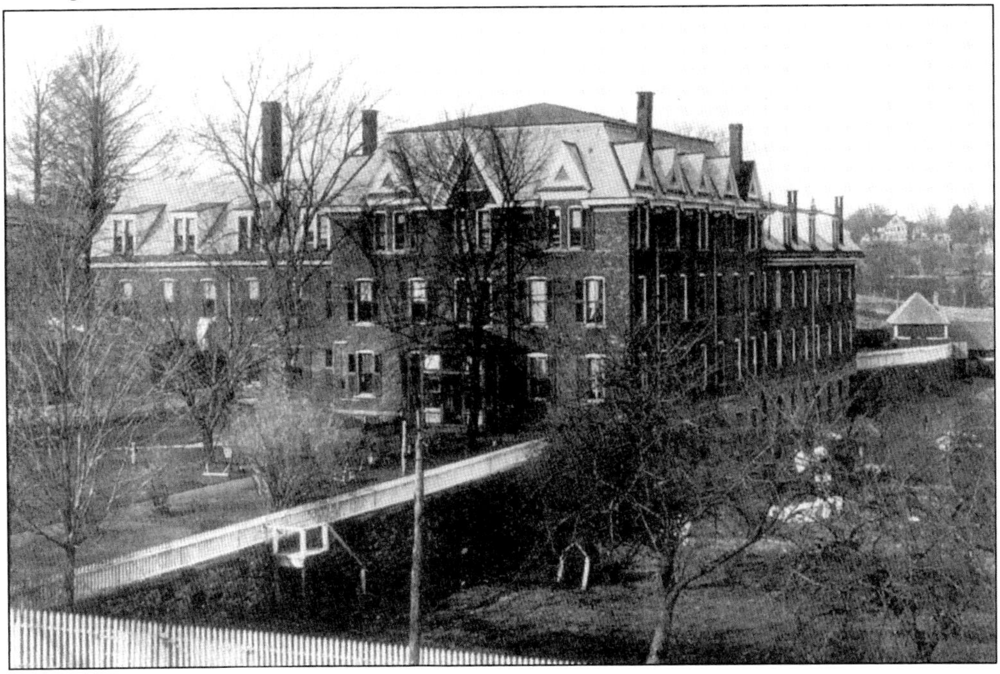

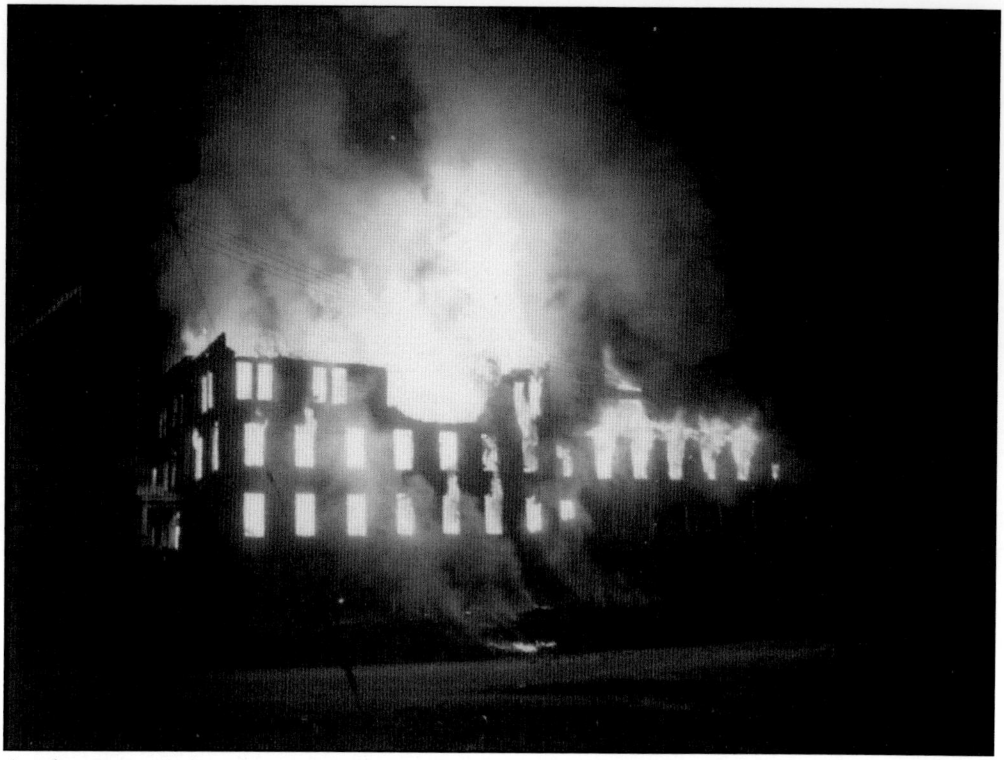

At the time of the alarm, significant fire spread had already occurred in the almshouse. Responding Norwich units were presented with a large conflagration.

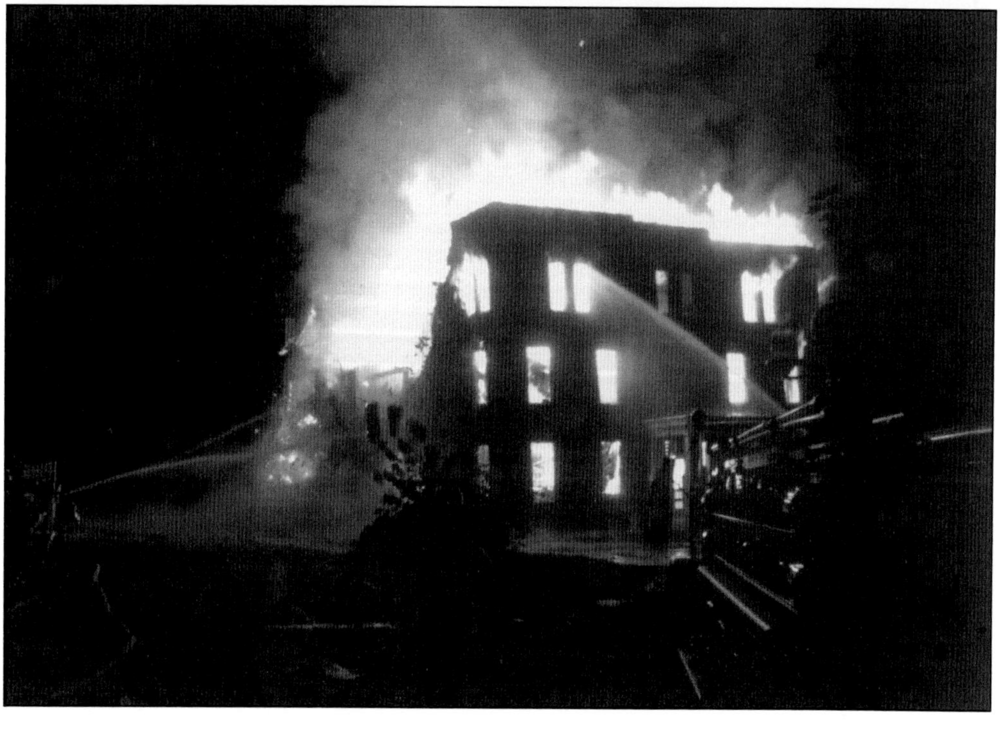

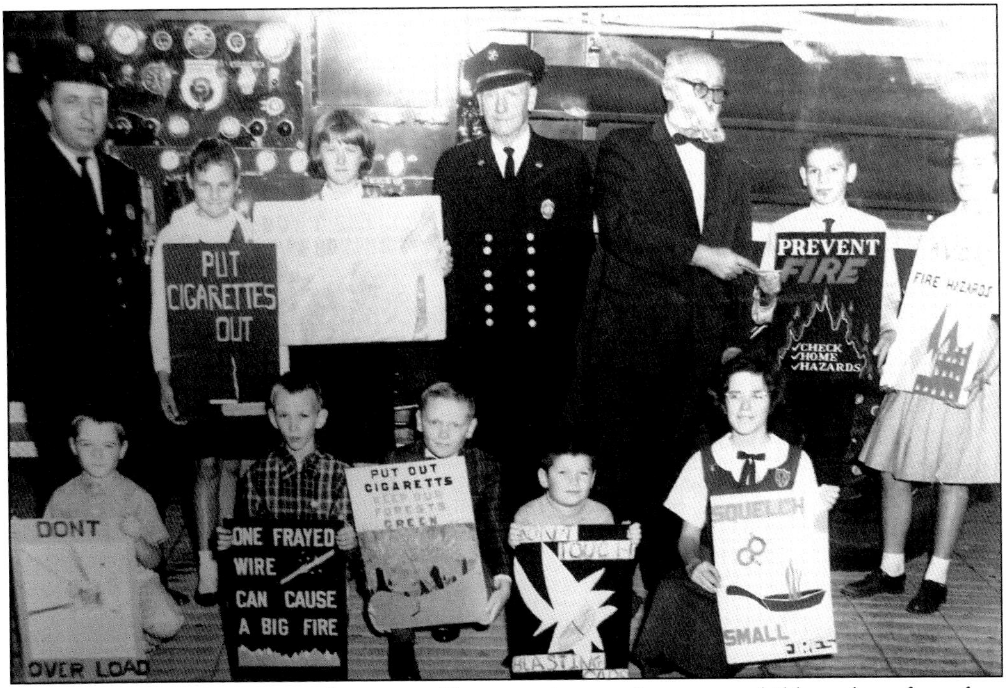

The Norwich Fire Department began a public campaign to educate area children about fire safety. From left to right, firefighter Harold Lamphere Jr., Chief William Confrey, and Henry Lucas (city council president) are pictured with the winners of the annual fire prevention poster contest.

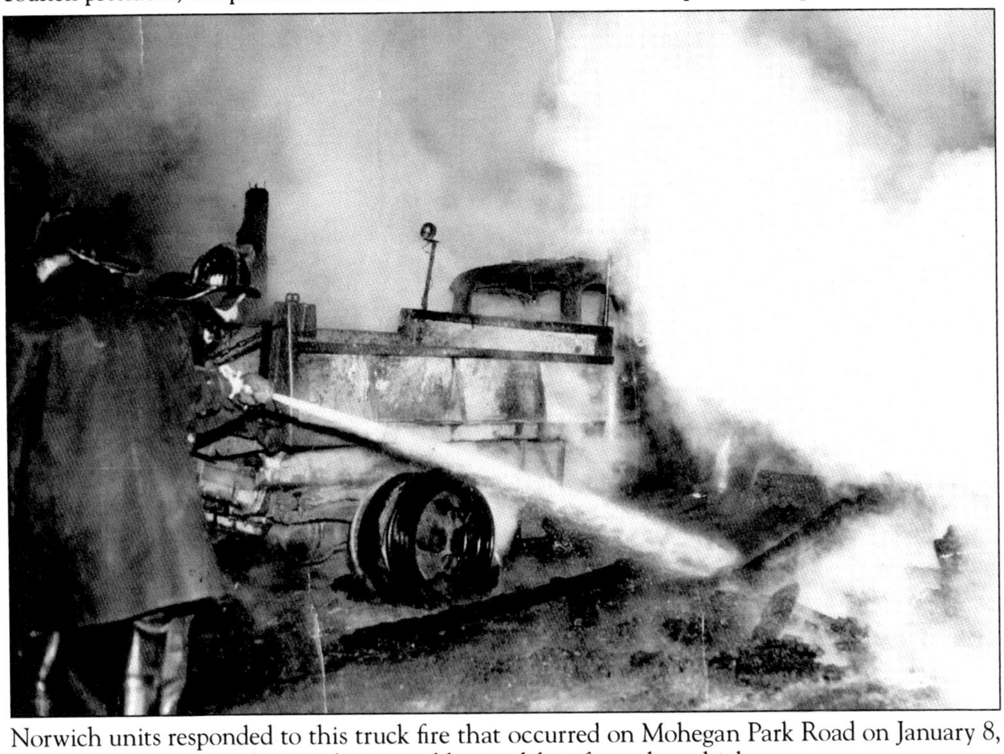

Norwich units responded to this truck fire that occurred on Mohegan Park Road on January 8, 1959. Firemen are seen hosing down smoldering debris from the vehicle.

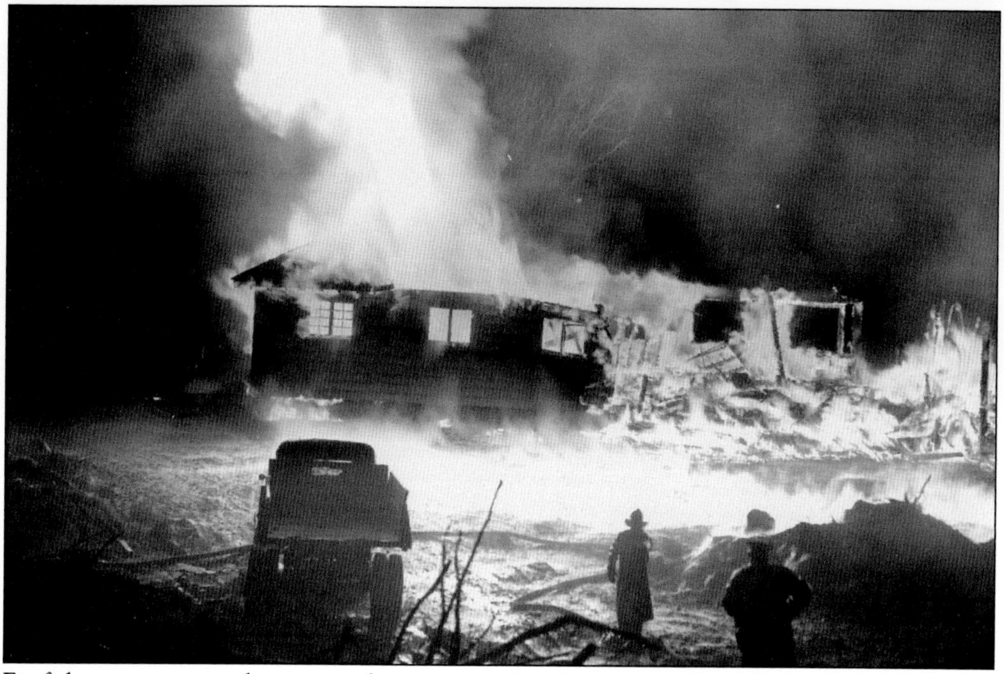

Firefighters arrive on the scene of a structure fire that occurred in the newly transplanted recreational center of SS. Peter and Paul Church. This building was moved from the Uncas on Thames Hospital to SS. Peter and Paul Church on Elizabeth Street. This 1950s fire was supposedly started when the fireplace was left unattended.

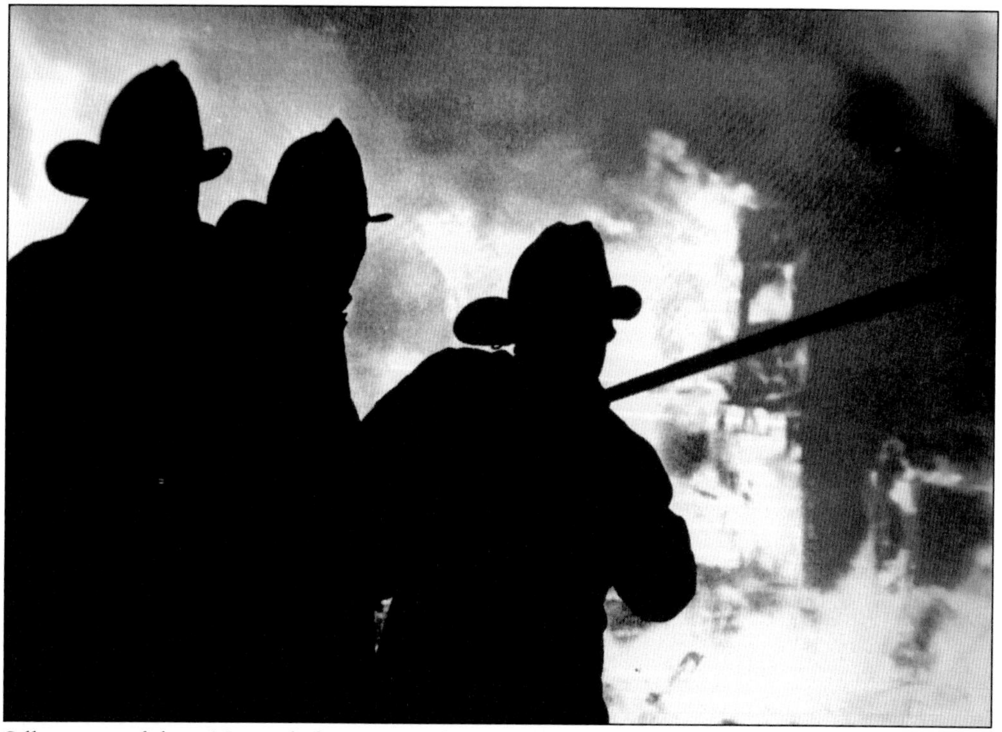

Silhouettes of three Norwich firemen are shown in contrast to the bright blaze.

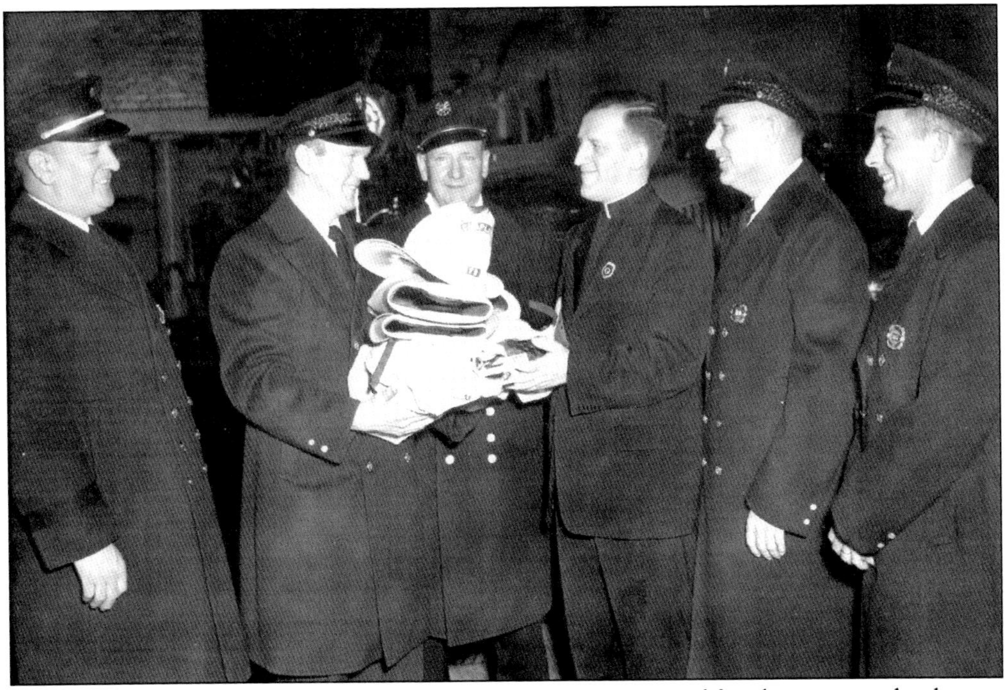

Fr. Henry Dziadosz of St. Joseph's Church accepts the position of fire department chaplain in this 1950s photograph. From left to right are Capt. Howard Ludlow, fireman William Dougherty, Chief William Confrey, Dziadosz, fireman Henry Milewski, and fireman Harold Lamphere Jr.

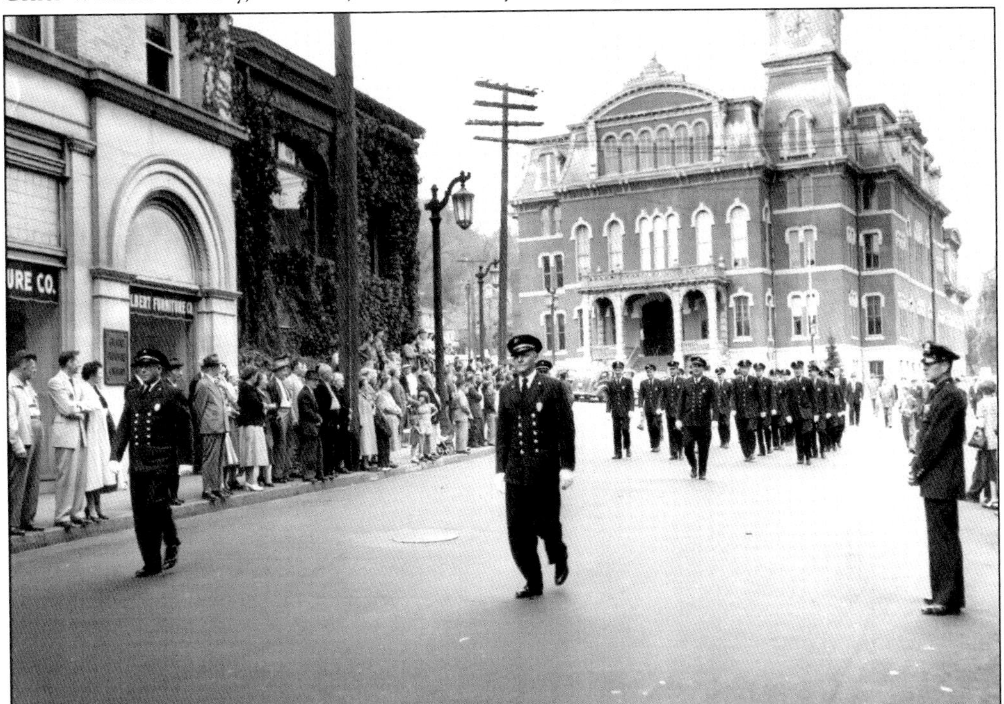

Department personnel are seen here participating in a parade through Union Square. Firefighters are led by Chief William Confrey (center) and Deputy Chief William LaRochelle (left).

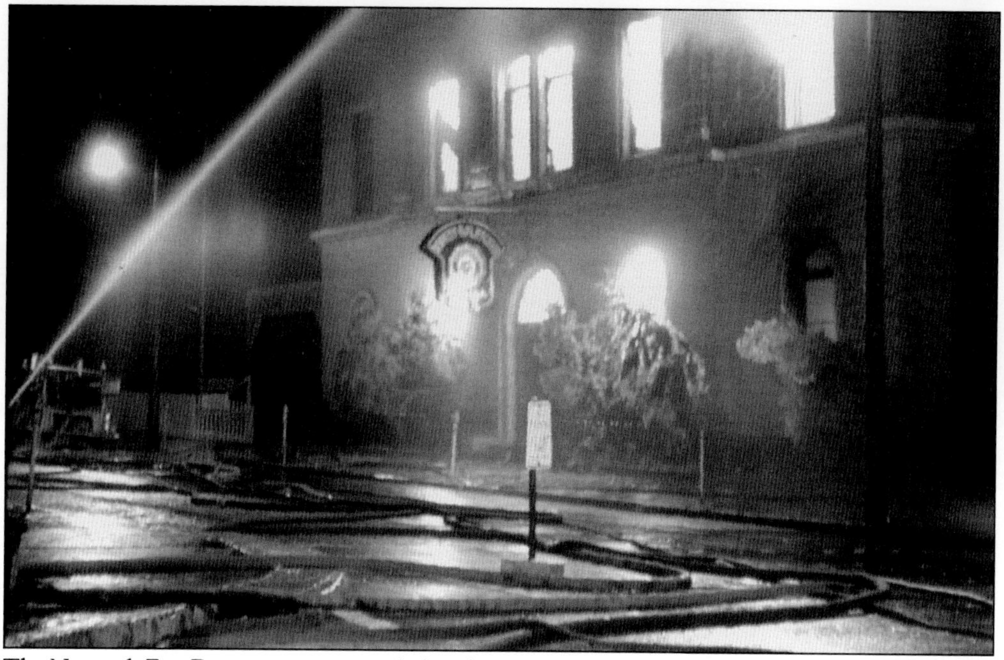

The Norwich Fire Department responded to the American Legion located at 95 Broadway on the morning of February 24, 1960. The large brick building, formally occupied by the Free Masons in the late 1800s and early 1900s, caught fire just after 3:30 a.m. The fire was believed to have been caused by a faulty refrigerator that ignited wood paneling. Below, Aerial Truck Company No. 1 is shown here operating a ladder pipe on the three-story structure.

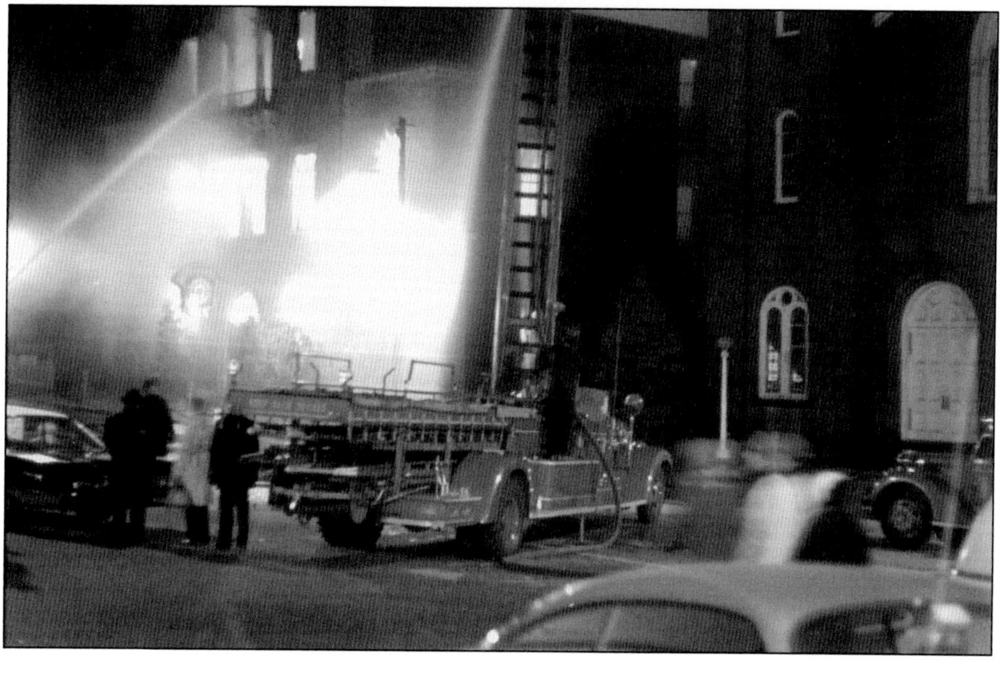

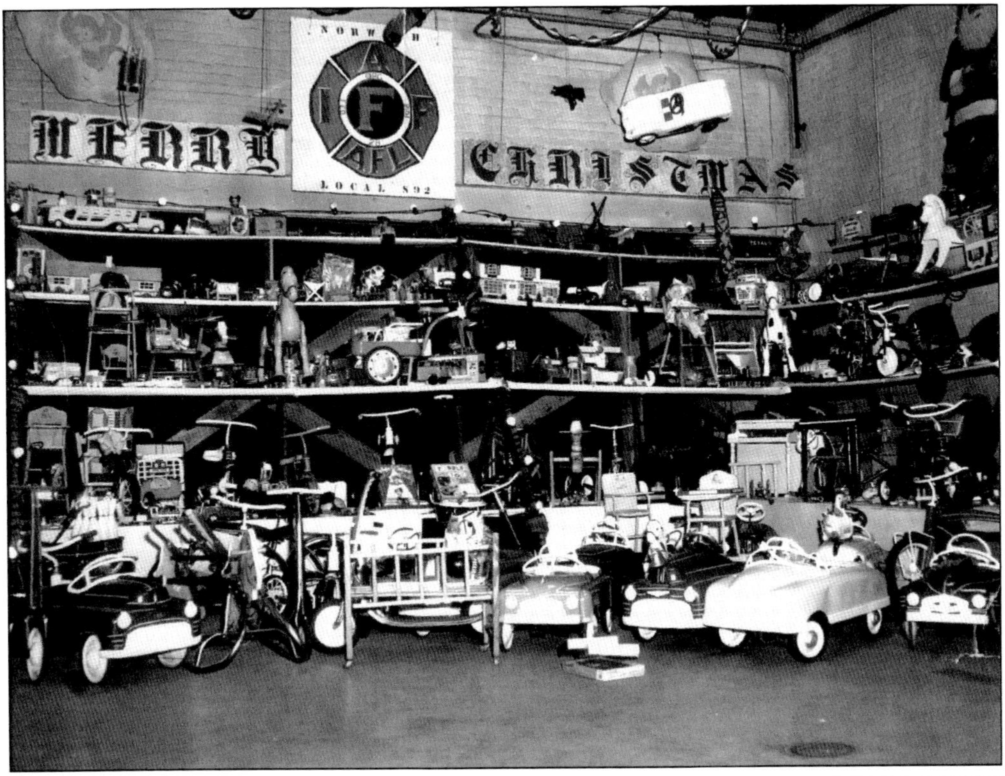

During the Christmas season, the Norwich Fire Department participated in a local toy drive that benefited the Salvation Army and the United Workers Group. Many toys were built and repaired by members of the Norwich Fire Department for the area children. At right, firefighters organize tricycles and wagons for the toy drive. Pictured from left to right are Edward Romano, Raymond Brosfske, Thomas Demauro, and Thomas LaFreniere.

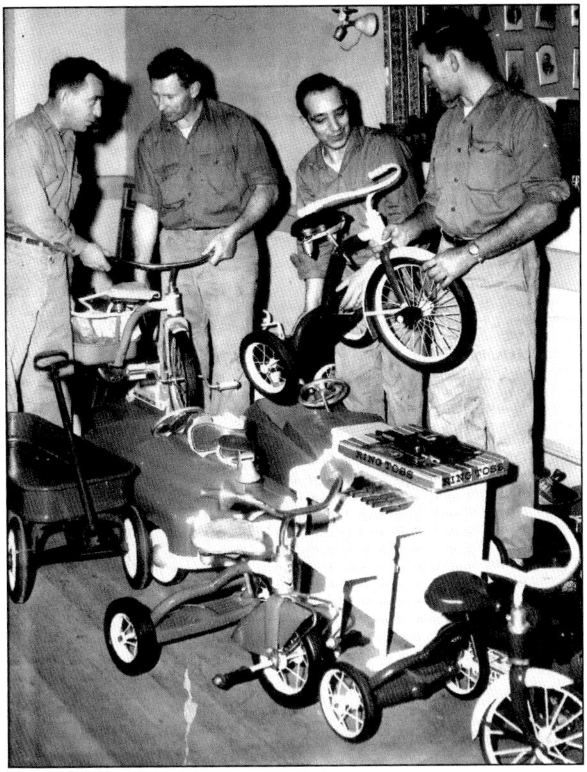

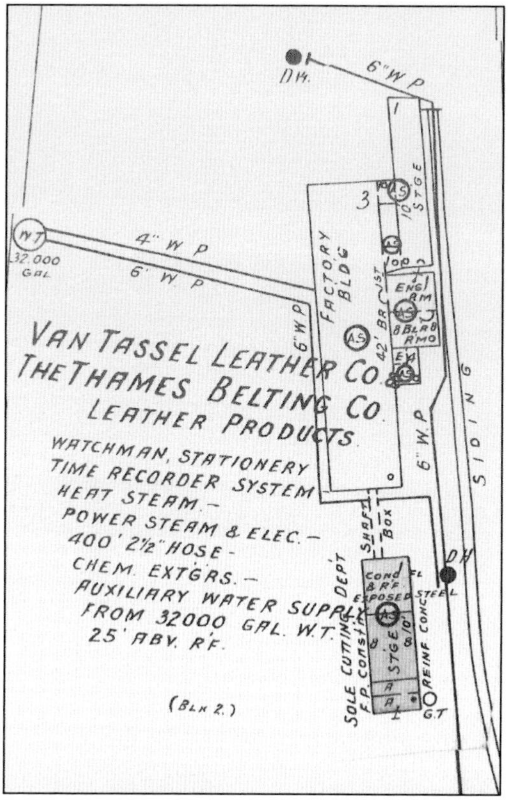

On April 3, 1962, the Norwich Fire Department lost four men in a violent and fiery explosion at the Van Tassel warehouse on Forest Street. Killed in the tragedy were Capt. William Sheridan, Leonard Counihan, Edward Romano, and Carl Burke. An employee reported smoke coming from the trailer of a truck that was being unloaded at the dock. The truck was carrying a 20-ton load of organic peroxides from New York that may have been ignited by friction or a leak. The alarm from box 125 was received at 1:25 p.m., and Norwich's two engines and the ladder truck were dispatched. Firefighters crouched behind the concrete loading platform wall and directed a two-and-a-half-inch hose stream into the rear of the smoldering trailer. The explosion that killed four firefighters and caused damage throughout a large portion of the city and was felt as far away as Montville and Preston. This tragedy led to stricter guidelines for transportation and placarding of hazardous materials for the entire country. The preceding passage was contributed by Pvt. Steven Fricke, Norwich Fire Department.

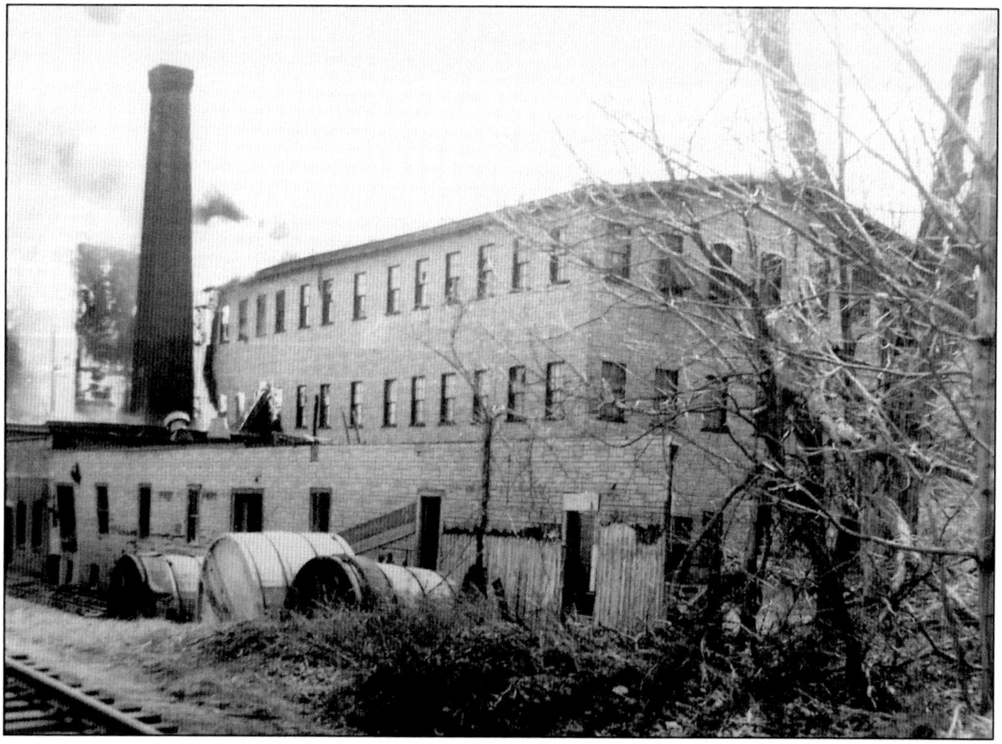

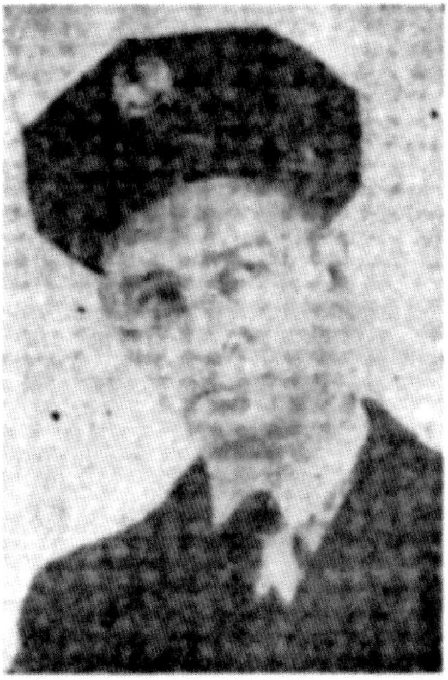

CAPT. SHERIDAN

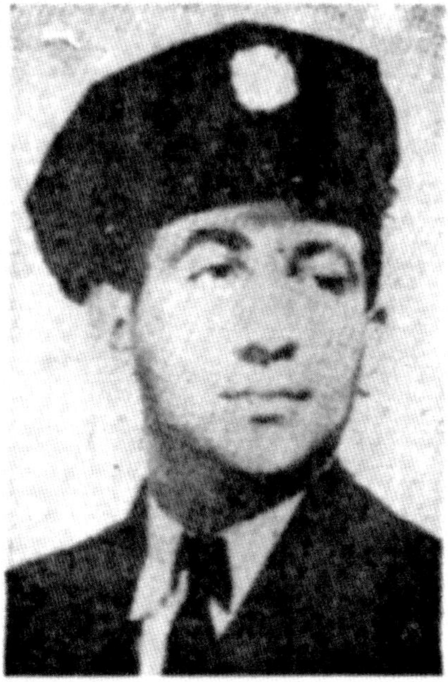

FIREMAN ROMANO

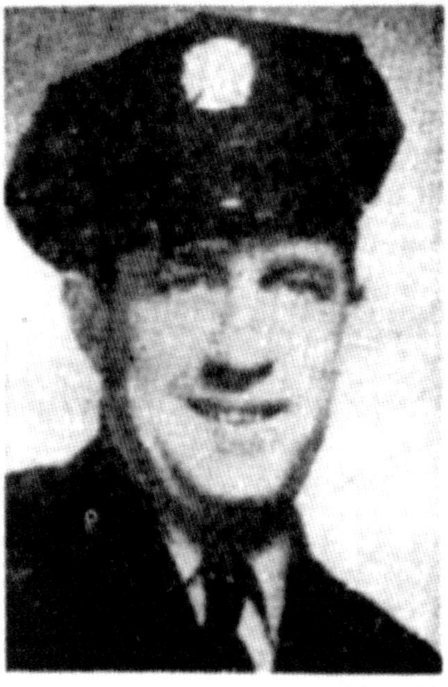

FIREMAN COUNIHAN

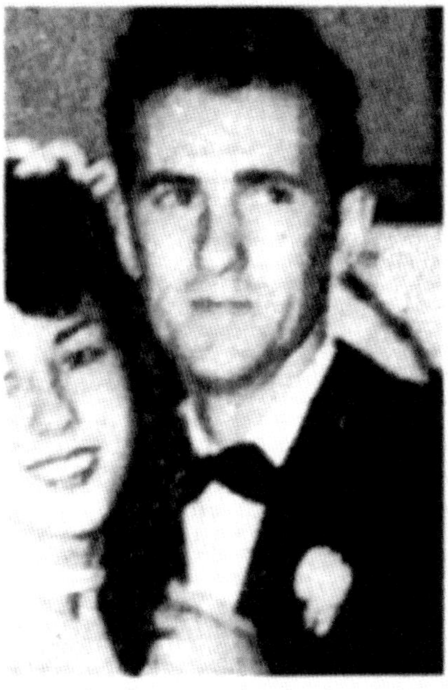

FIREMAN BURKE

On April 3, 1962, the Norwich Fire Department suffered its worst loss with the death of four men. Capt. William Sheridan (24 years), Leonard Counihan (15 years), Edward Romano (14 years), and Carl Burke (3 years) perished in a fiery explosion at the Van Tassel warehouse on Forest Street.

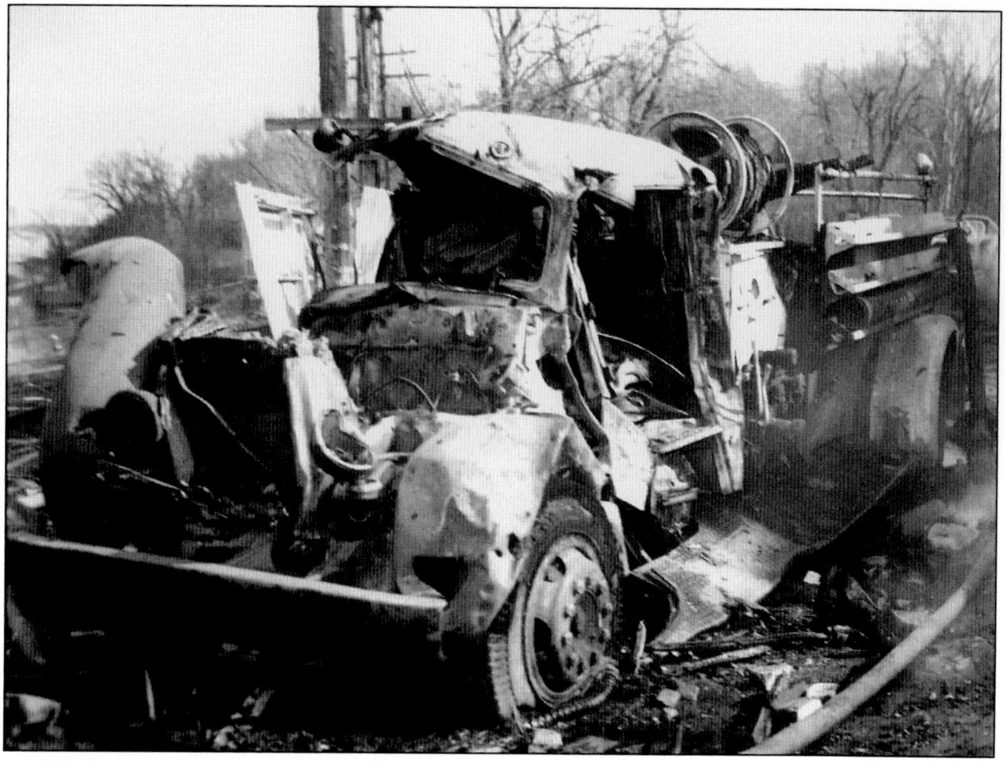

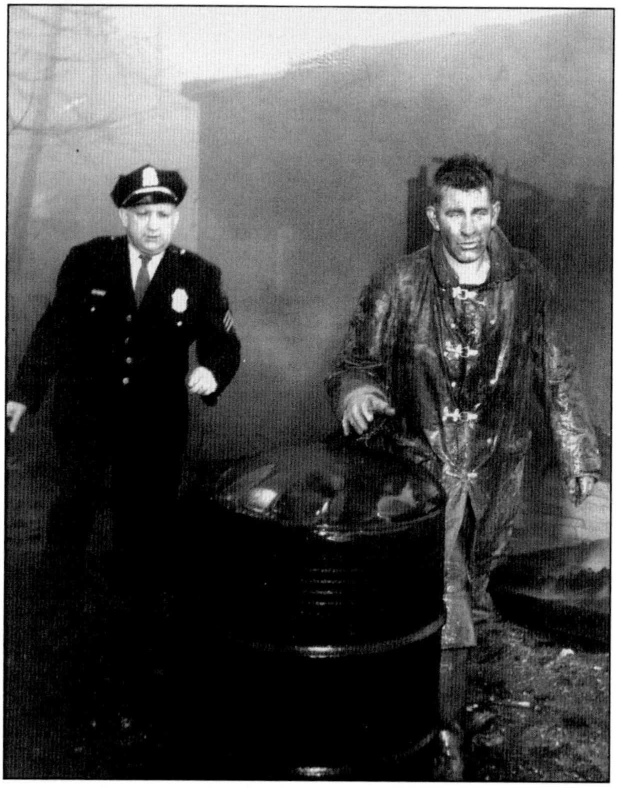

Above, twisted steel is all that remains of this engine. At left, Lt. Thomas LaFreniere, one of three survivors of the explosion, is seen walking back to his fallen men. Police sergeant John Sisco is seen pursuing LaFreniere. The extreme heat of the blaze is evident in the expanded 55-gallon drum.

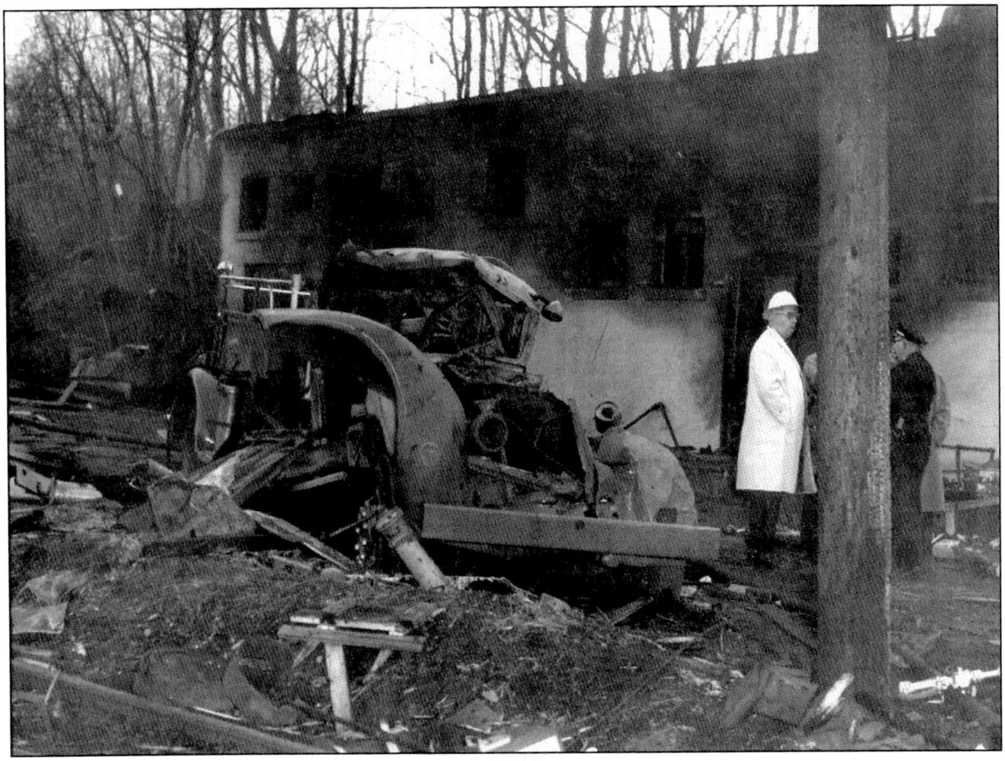

Engines No. 1 and No. 3 lie among the debris field.

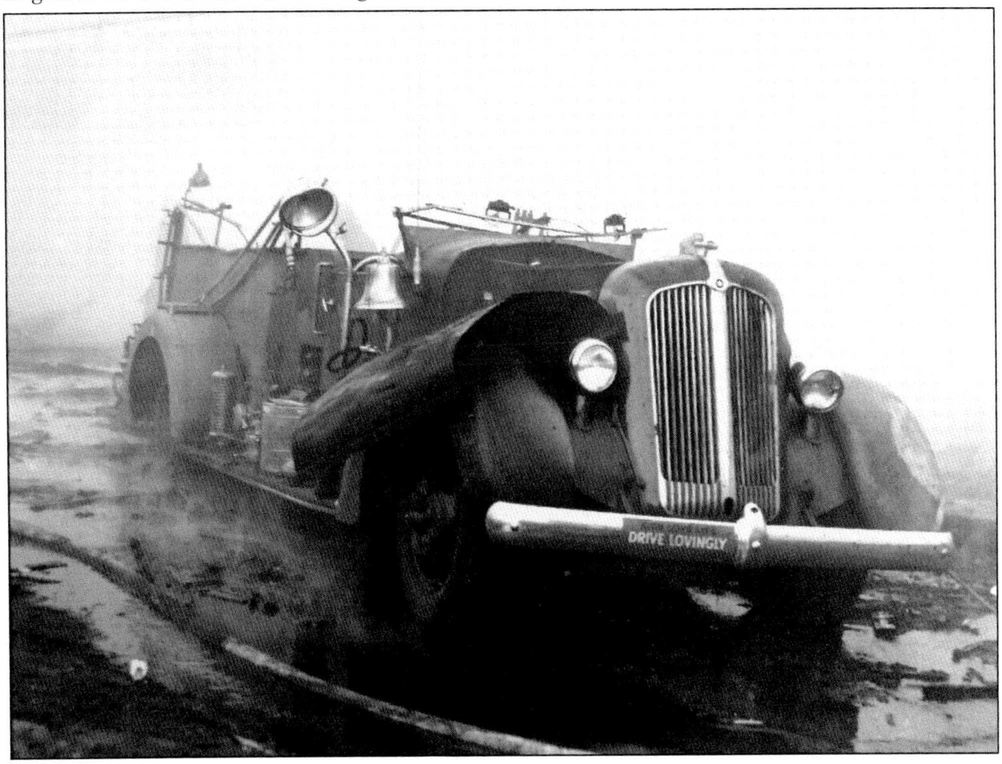

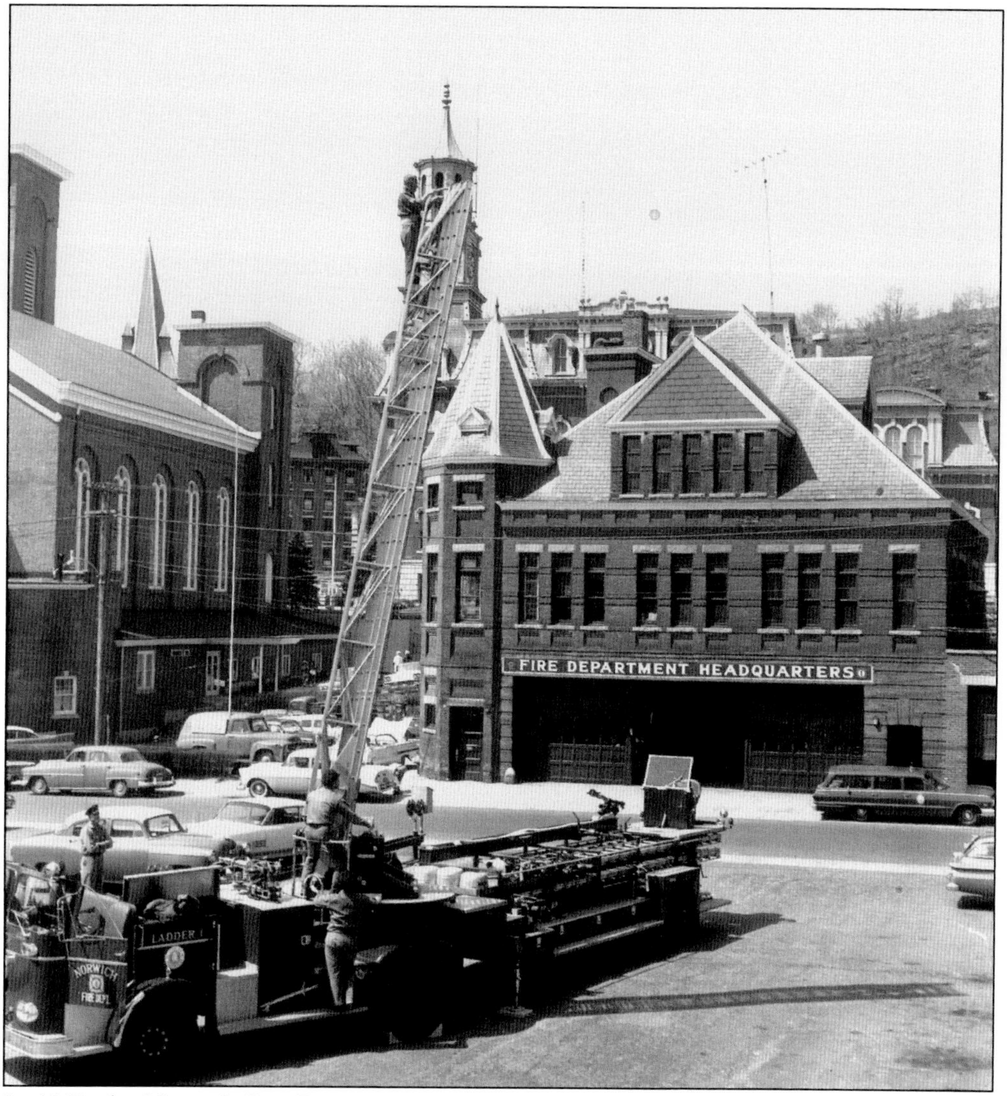

In 1963, the Norwich Fire Department purchased a new 100-foot Seagrave tiller ladder for $42,000. The new 100-foot aerial designated ladder No. 1 was stationed at 24 Chestnut Street (headquarters) and replaced the 65-foot mid-mount ladder purchased in the late 1940s. The 1940s-model ladder was relocated to the Greeneville station as a reserve. Ladder No. 1 is pictured outside fire department headquarters performing an operational check. Firefighter Don Cilley is pictured on the ladder, firefighter Cal Cobb is at the controls, firefighter Ernest Champagne is located on the running board, and Capt. Edward J. Breen observes his men from the parking lot.

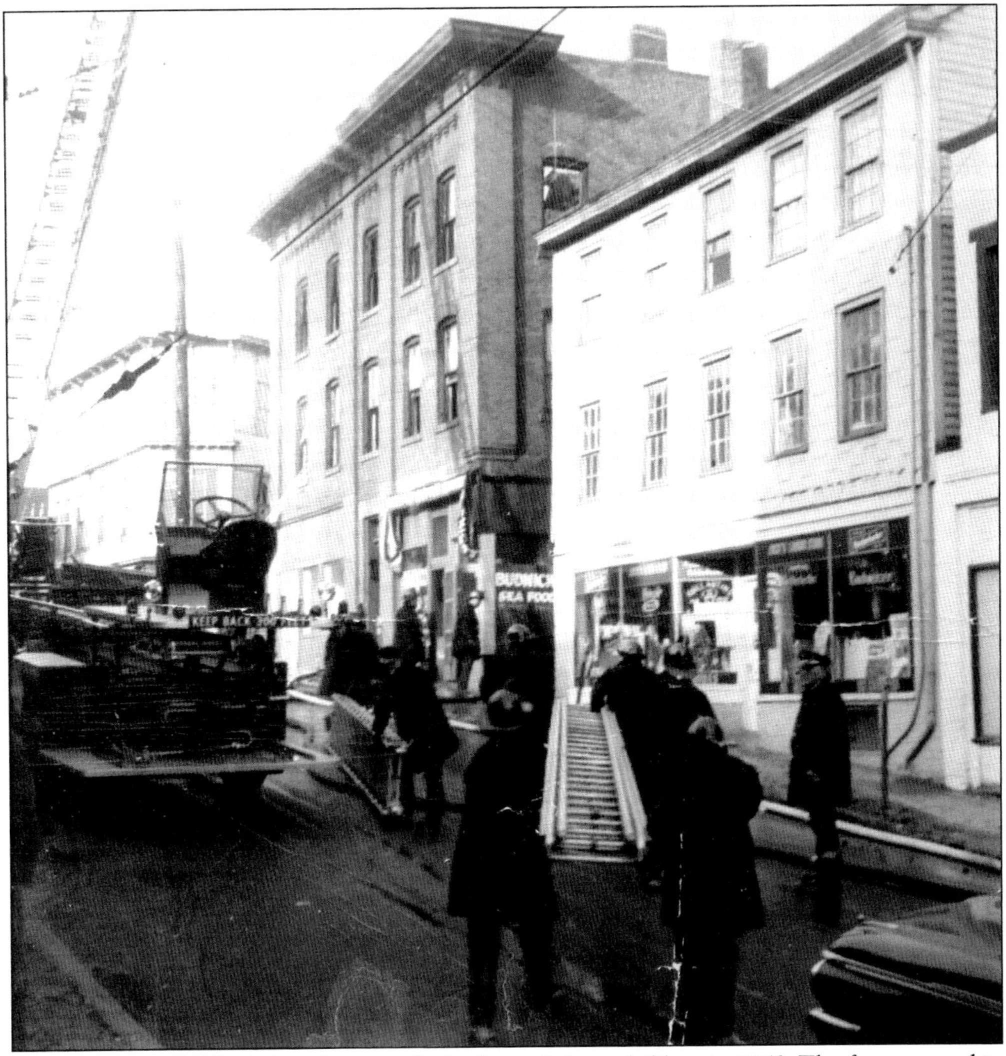

A fatal fire at 165 West Main Street took the lives of three children in 1963. The fire occurred in the upper apartment above Budnicks Seafood. The large three-story brick building was located on the corner of West Main and North Thames Streets.

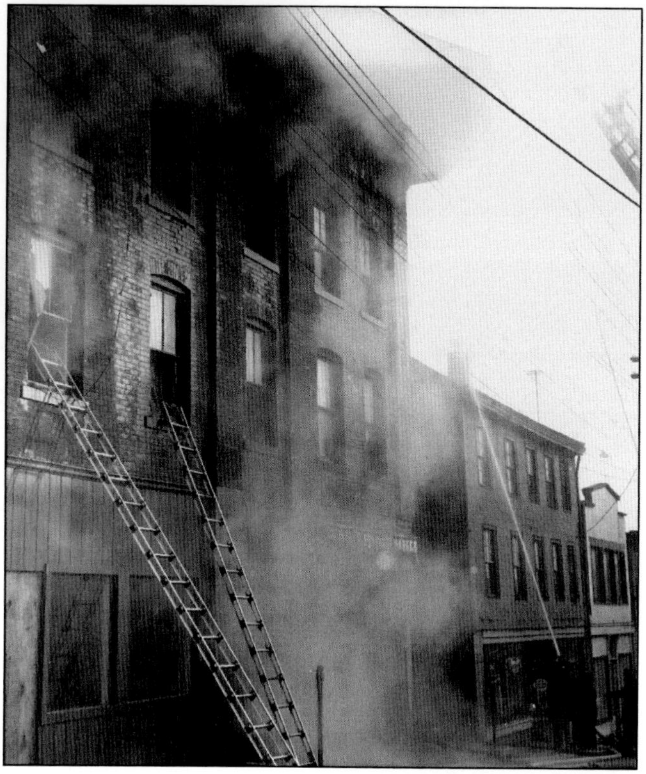

Above, firefighters direct hose streams under power lines in an attempt to wet down the eaves of 165 West Main Street. Below, a firefighter helps stretch three-inch hose from the hydrant to the engine.

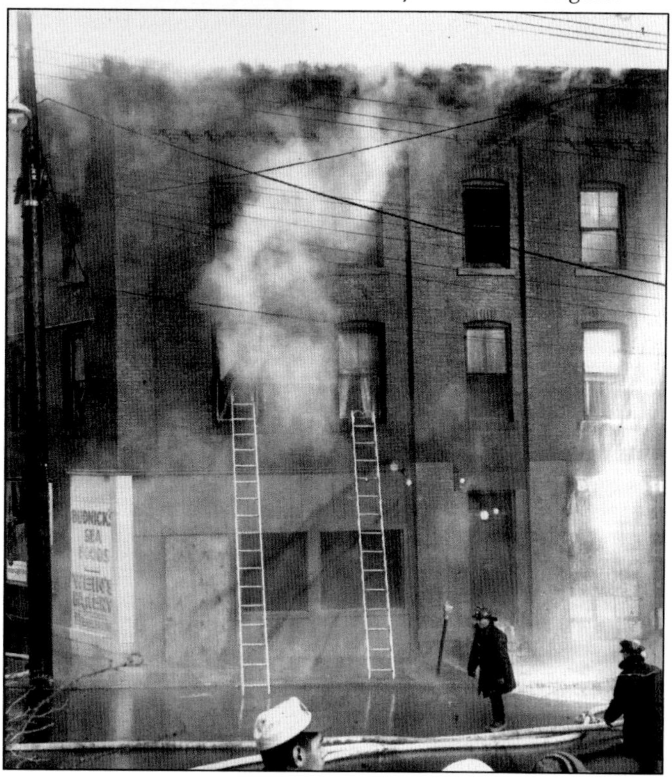

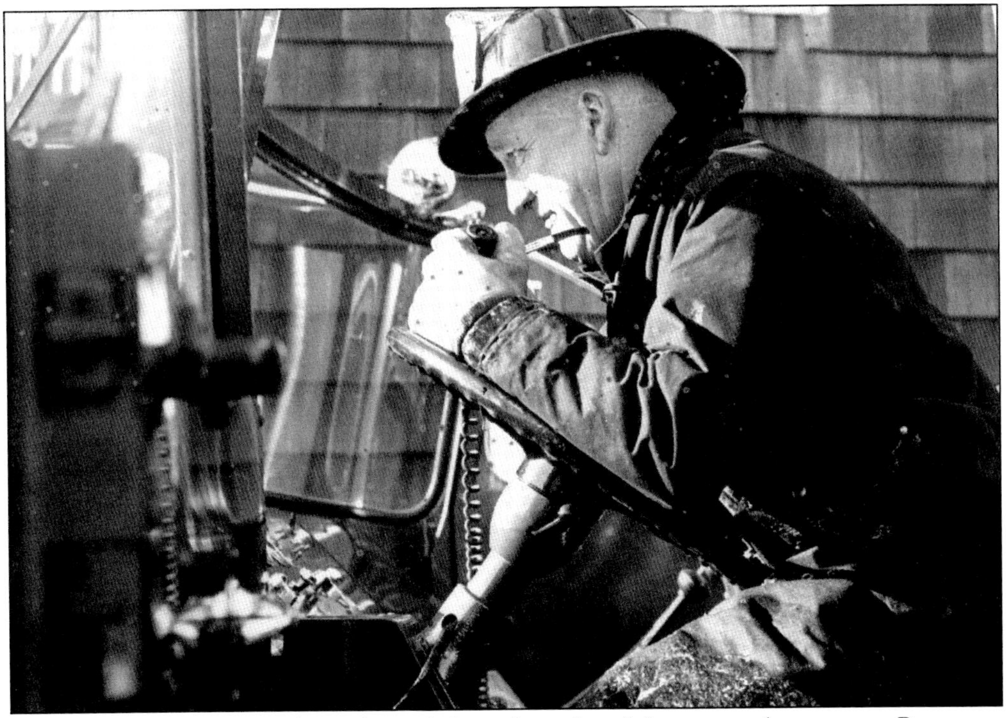

Capt. Edward J. Breen speaks to dispatch from the radio of this open-cab apparatus. Breen was seldom seen on the fire ground without his pipe.

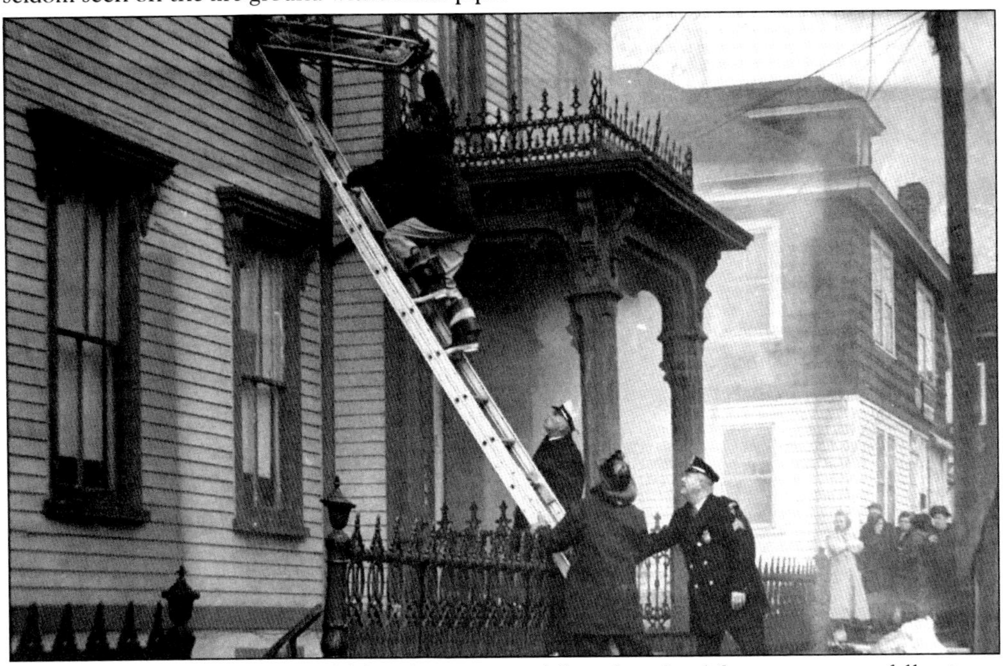

In this photograph, a 77-year-old female is removed from her third-floor apartment following a fire at 64 Boswell Avenue. The fire, which started in the basement of the building, was deemed suspicious. Chief Henry Milewski looks on from the ground as unidentified firefighters remove the patient.

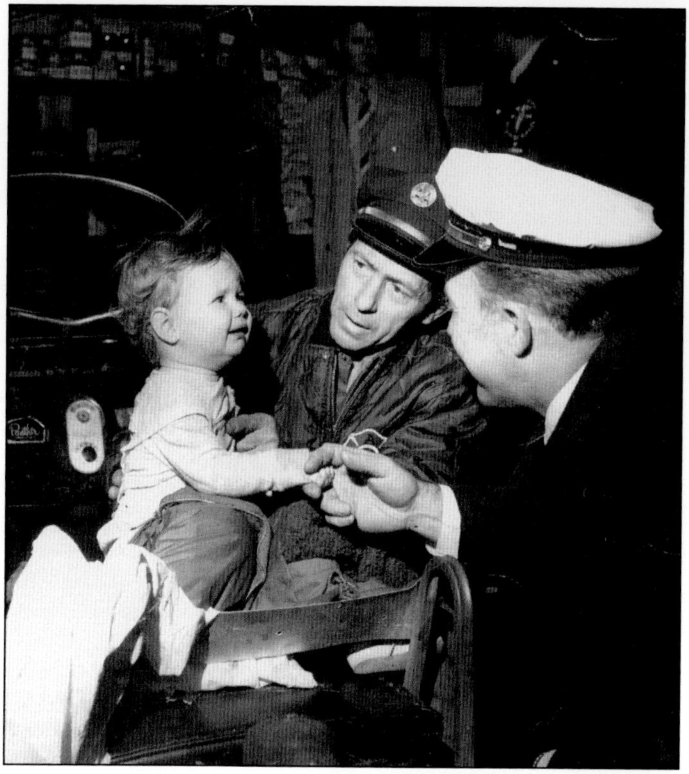

Capt. Cal Cobb and Deputy Chief Joseph Boisclair are seen here comforting a distraught child.

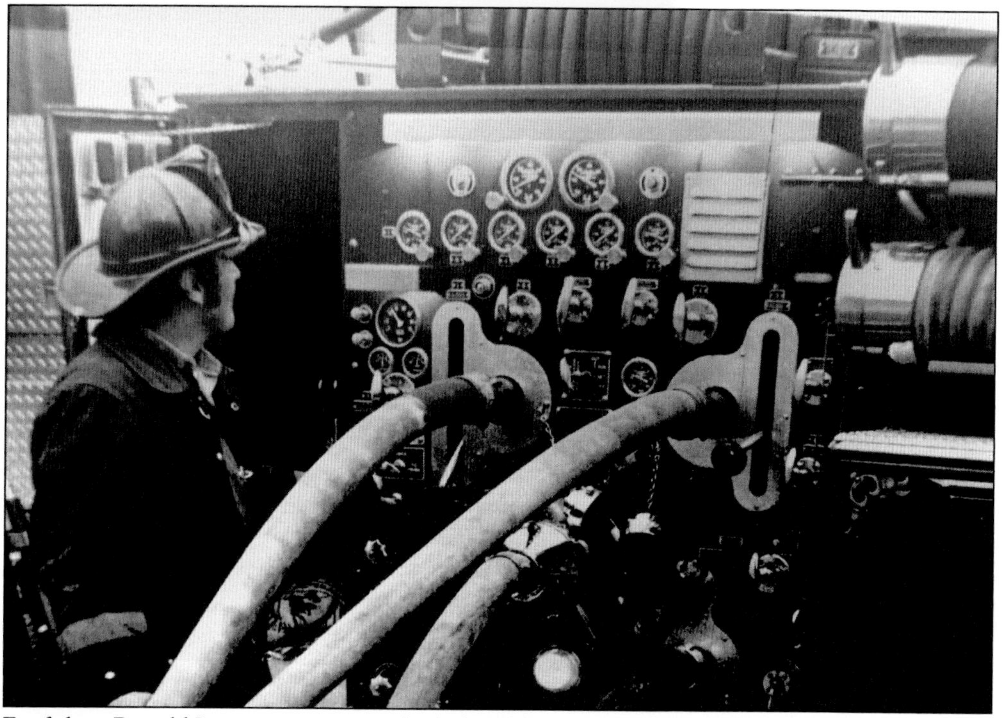

Firefighter Ronald Levine manages multiple hand lines at the controlled burn of Chappell Coal in 1969.

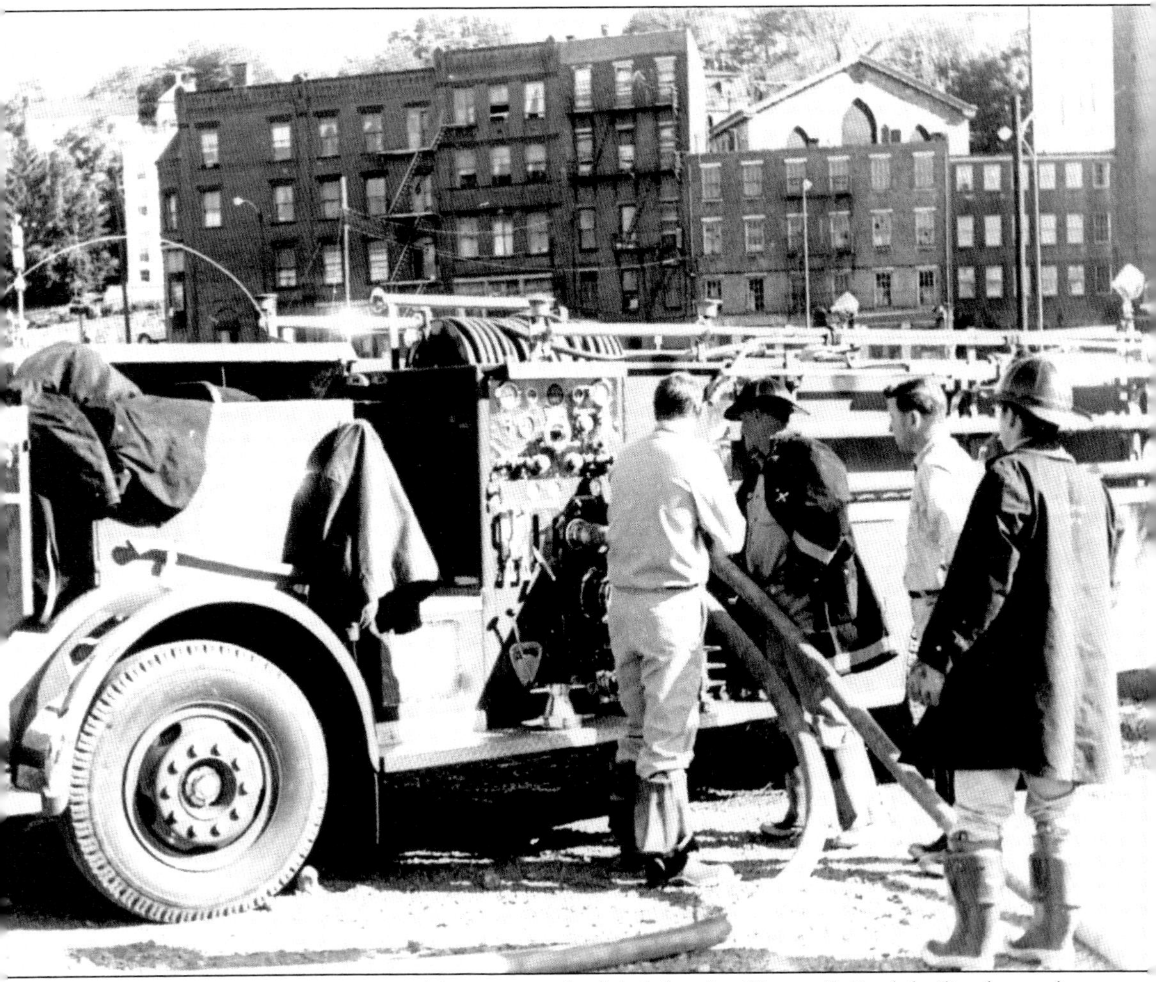

On September 21, 1969, a control burn was scheduled for the Chappell Coal facility located on Central Wharf. Four Norwich firemen are pictured adjacent to a 1962 Seagrave open-cab engine. This apparatus was purchased new in 1962 following the loss of equipment and four firemen in the Van Tassel warehouse fire. This Seagrave engine was capable of 1,000 gallons per minute and cost $23,500.

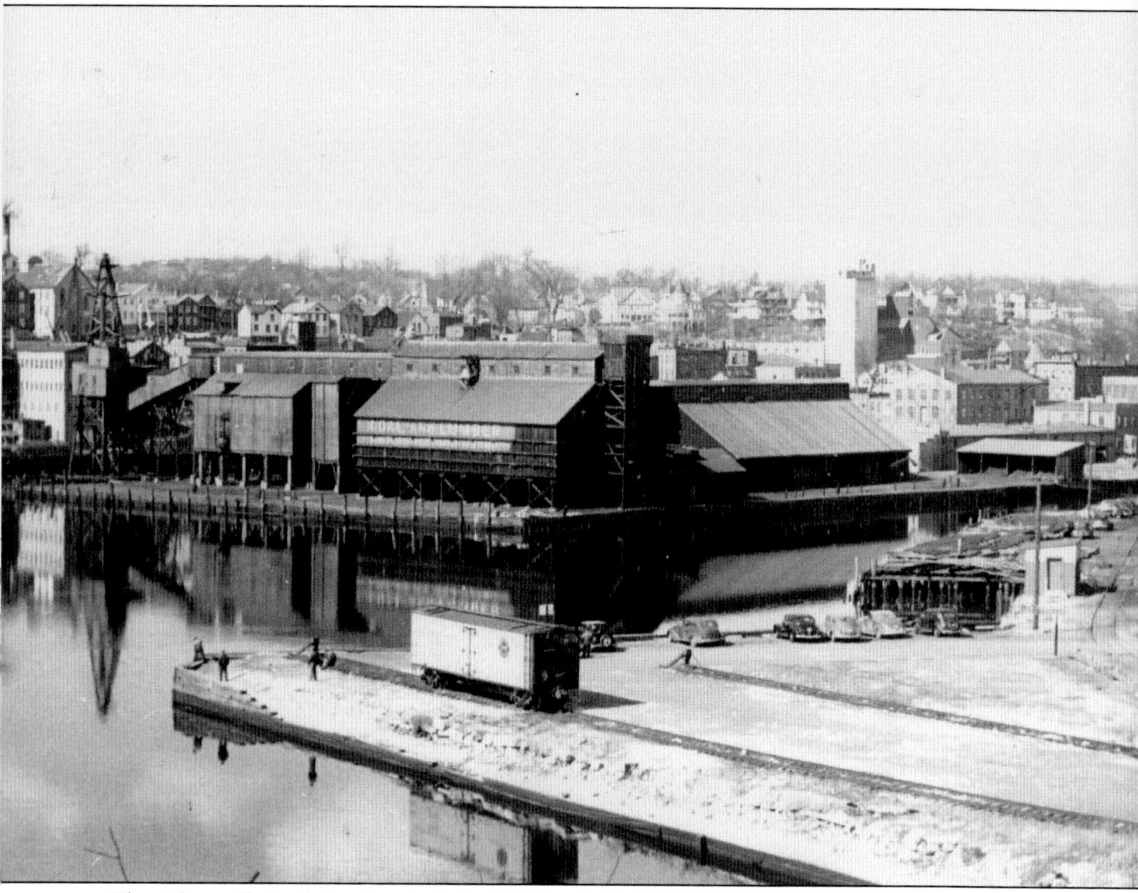

The Chappell Coal Company is pictured here in this *c.* 1940 photograph. This landmark had been around since the 1800s and employed many citizens throughout the years. The large wooden structures that occupied the southern tip of the wharf were demolished in 1969 for redevelopment.

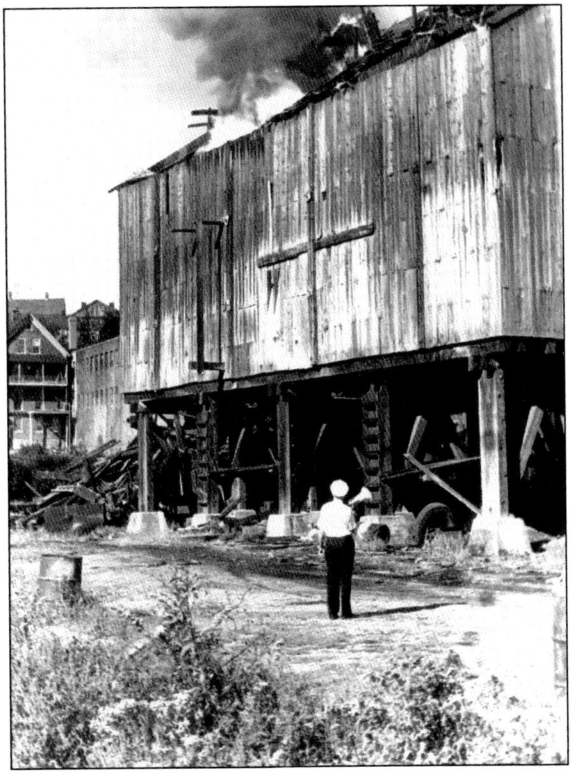

Chief Henry Milewski directs the controlled burn of the Chappell Coal Company on September 21, 1969.

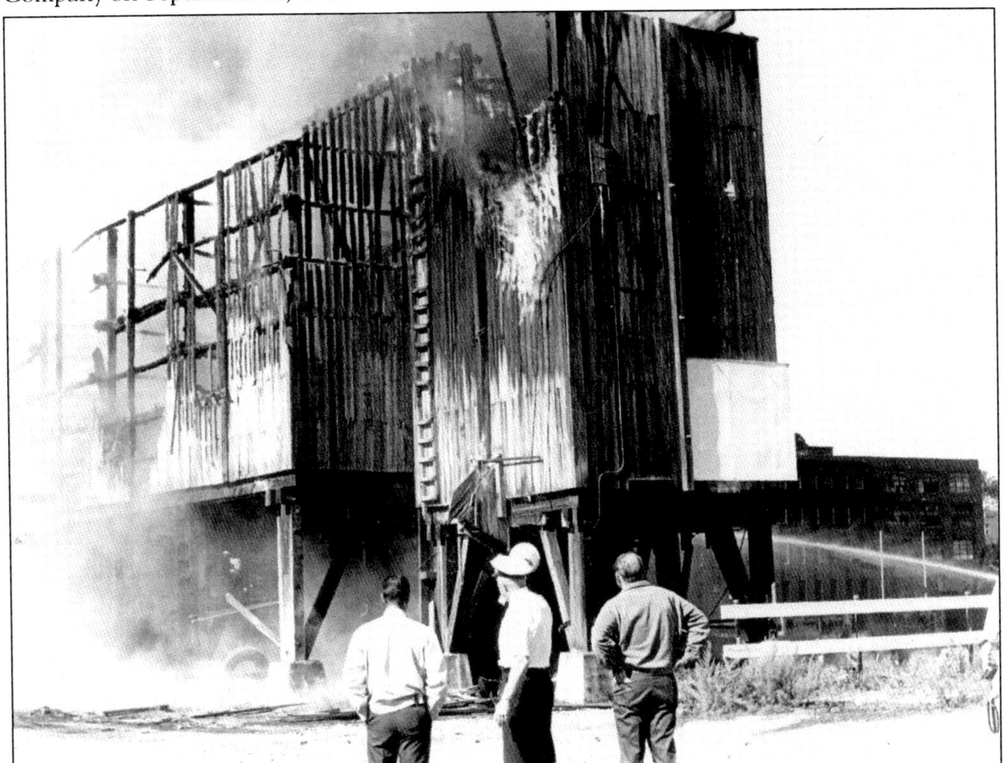

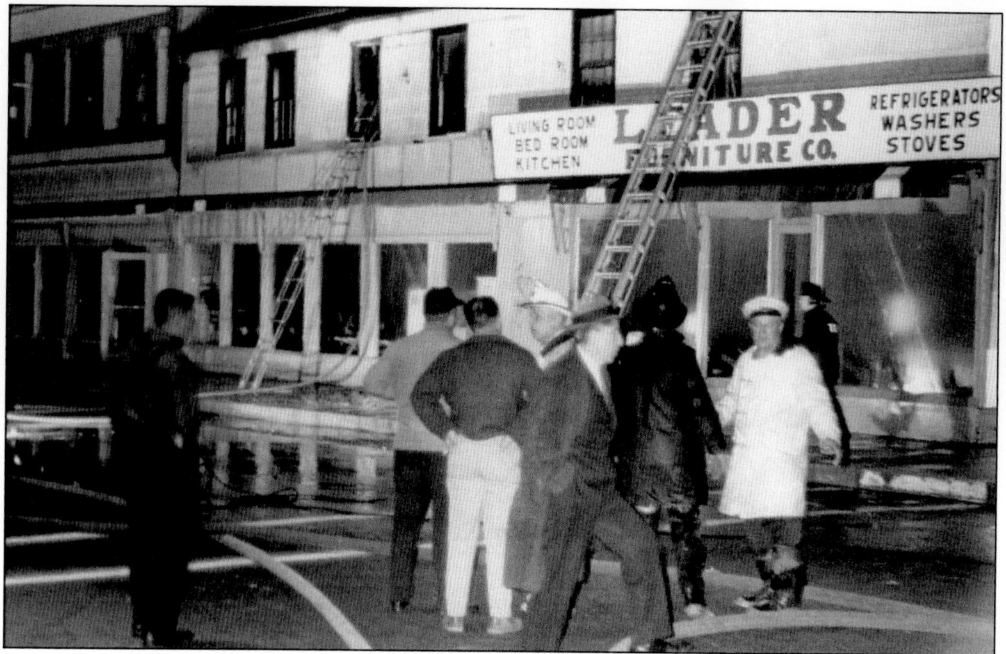

The Greeneville landmark Leader Furniture Company was struck by fire in 1969. Civilians as well as firefighters are seen walking among the fire hose. Chief Henry Milewski, pictured in the center of the photograph, speaks to two civilians as Deputy Chief Alexander Chmielewski directs the firemen.

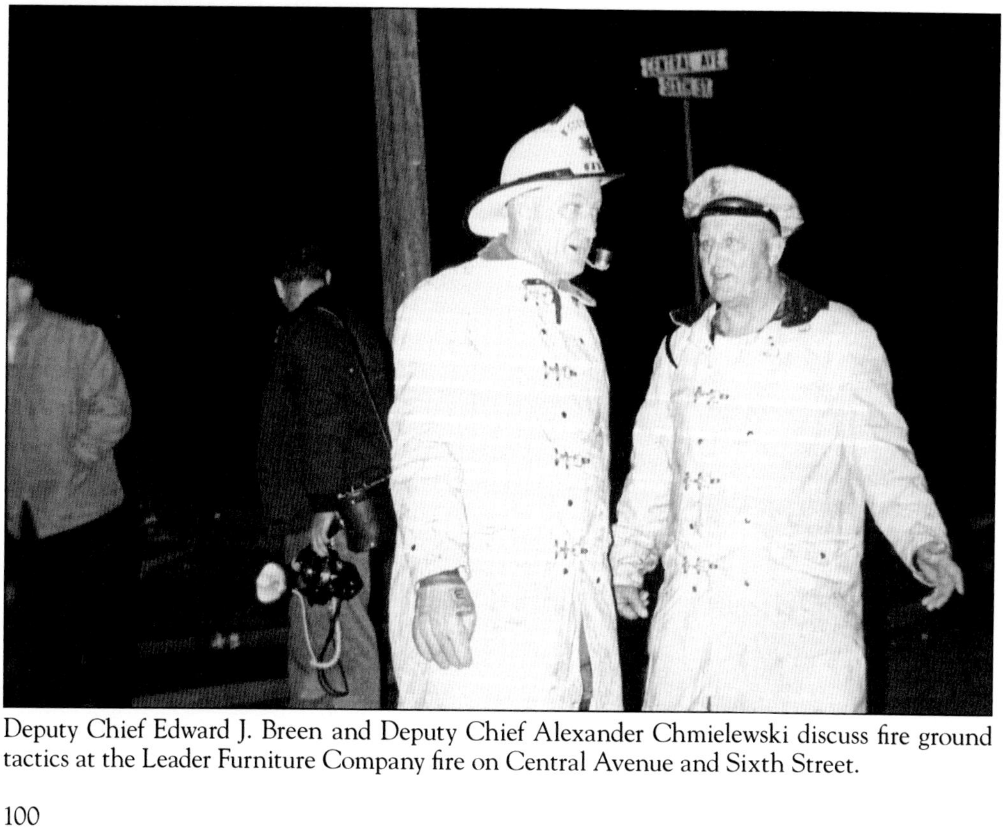

Deputy Chief Edward J. Breen and Deputy Chief Alexander Chmielewski discuss fire ground tactics at the Leader Furniture Company fire on Central Avenue and Sixth Street.

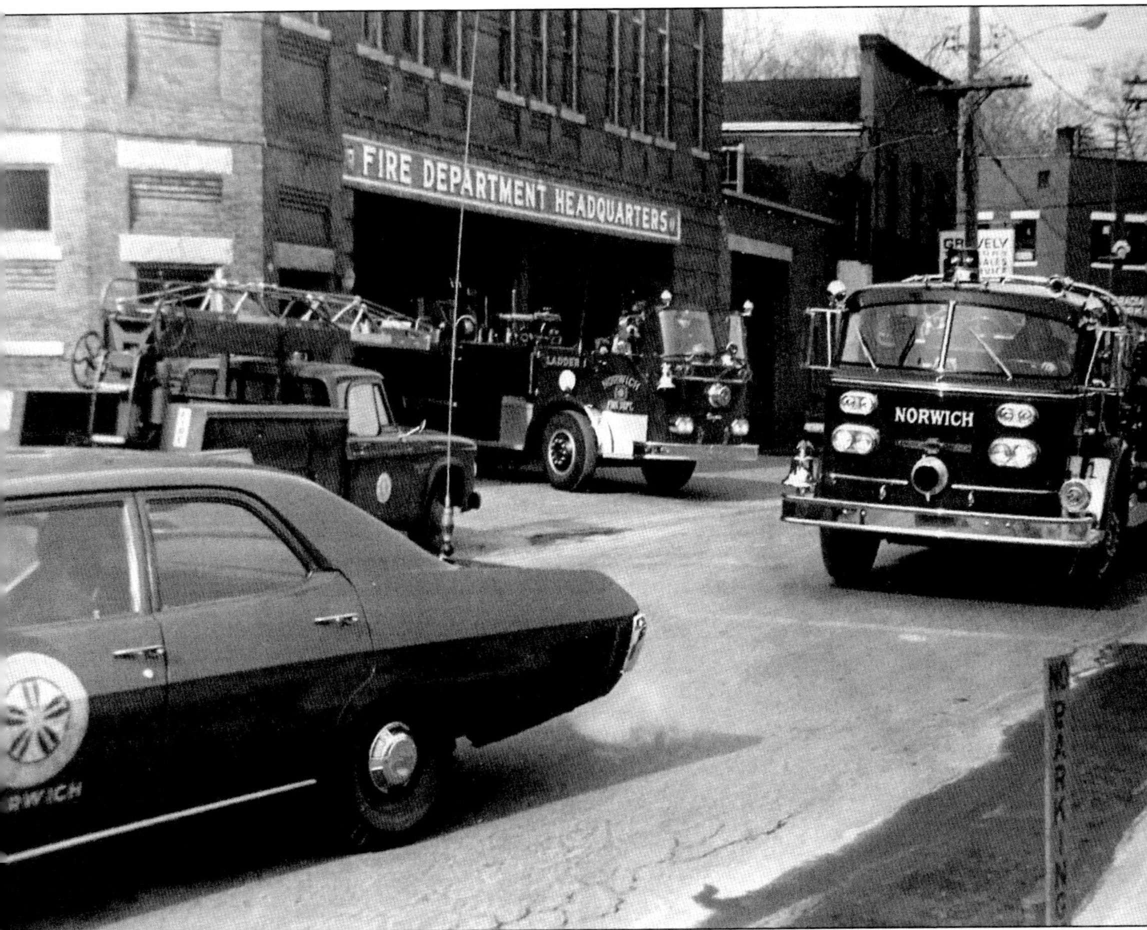

Chief Edward J. Breen responds to a fire alarm with squad A and ladder No. 1 in this 1971 photograph. The fourth vehicle in the photograph above is the fire alarm division's service truck. The Norwich Fire Department maintained and managed the city's fire alarm system from 1876 to the 1980s.

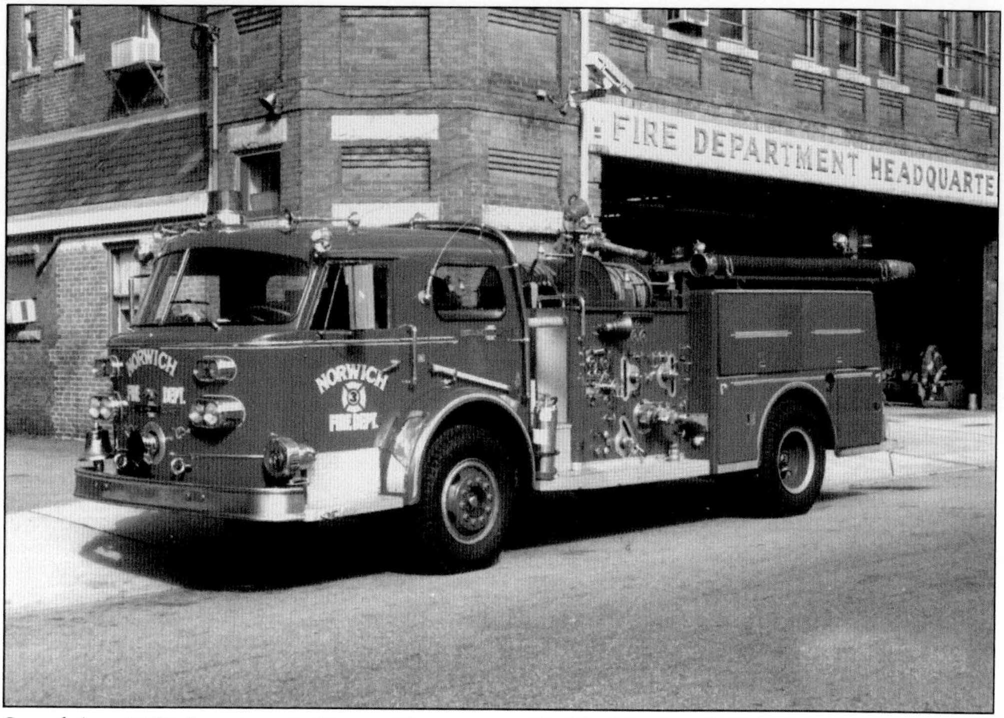

Squad A, a 1969 American LaFrance Century, and ladder No. 1, a 1963 Seagrave Tiller ladder, from page 101 are shown here in 1994 as engine No. 3 and ladder No 2.

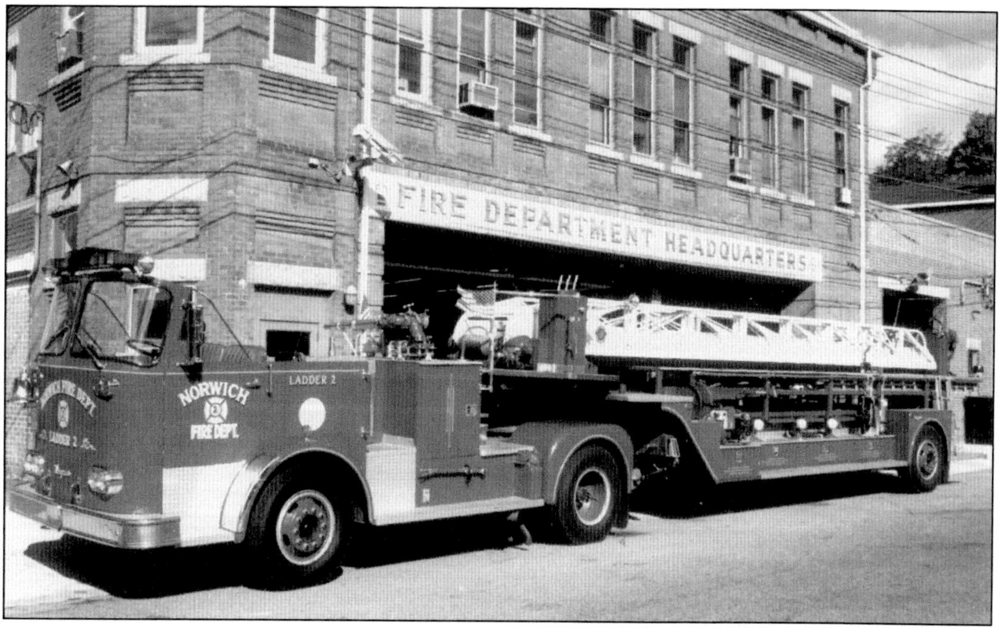

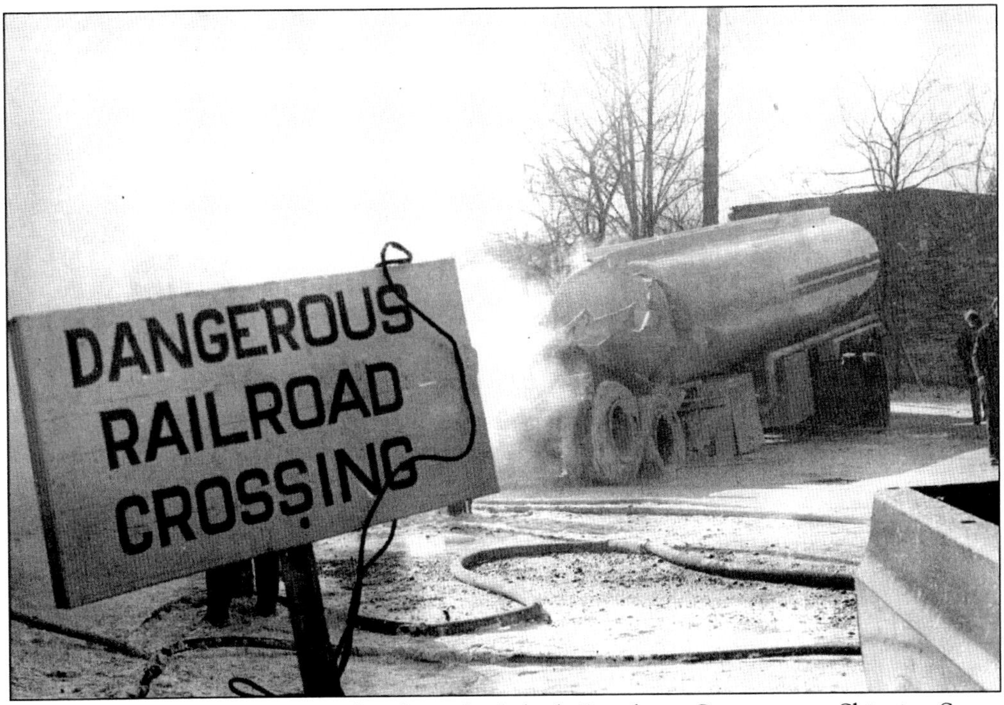

On June 15, 1974, a gasoline tanker from the Lehigh Petroleum Company on Shipping Street was struck by a locomotive from the Vermont Central Railroad. The fire that ensued caused considerable damage. Below, firemen apply foam to the flammable liquid in the tanker.

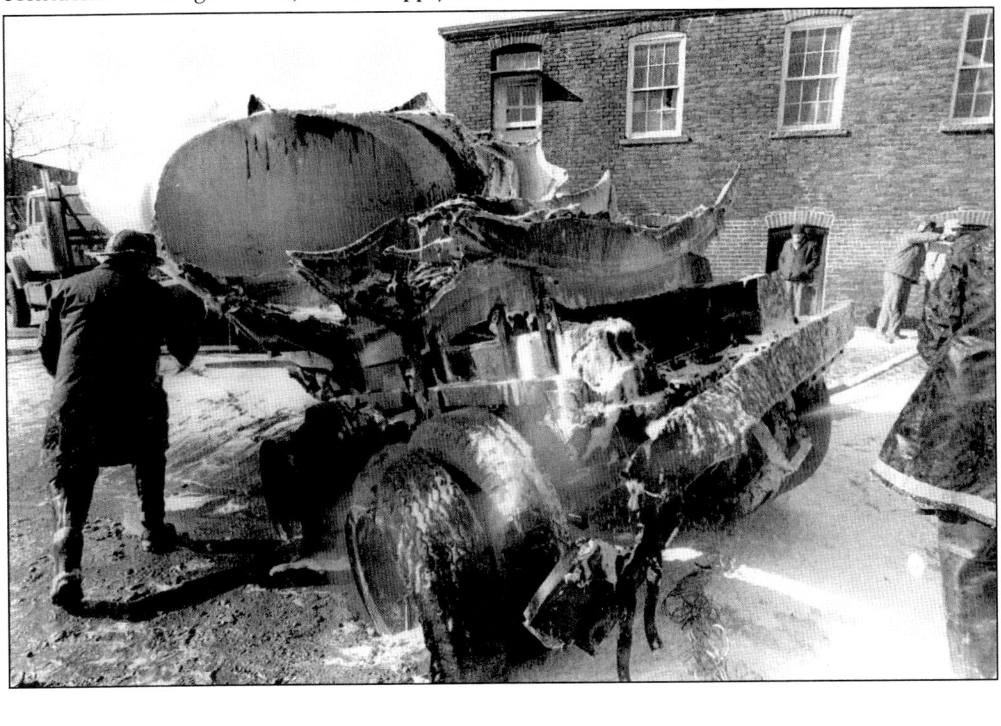

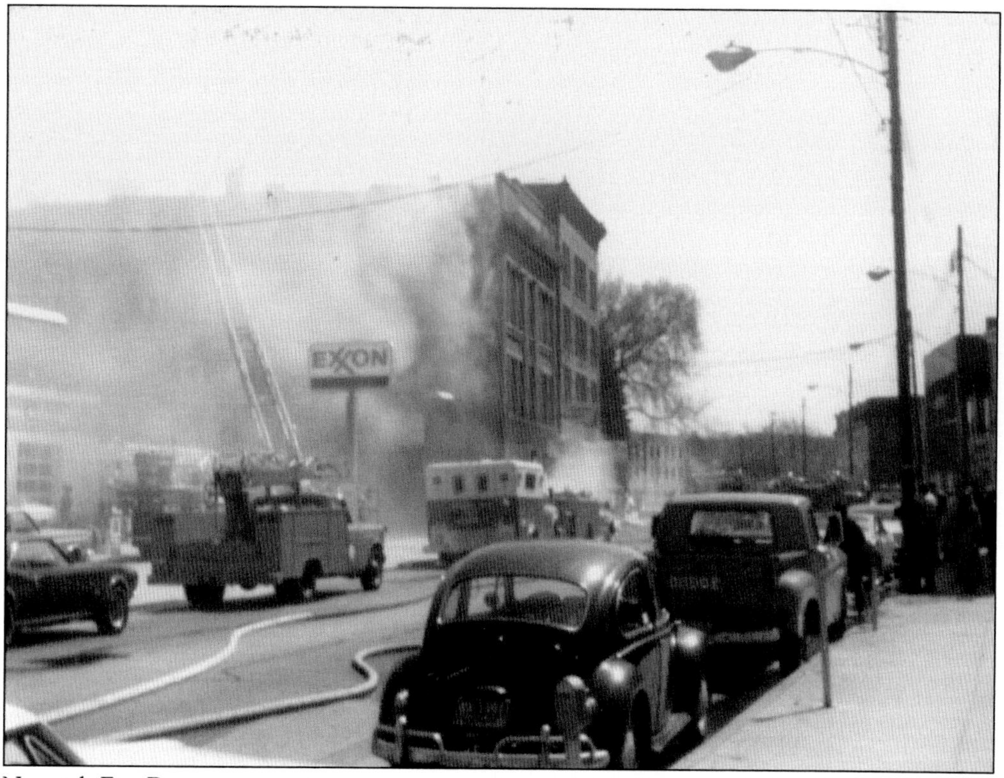

Norwich Fire Department units are seen operating on Main Street in the building occupied by the Main Street Café. The numerous partitions of the basement allowed for fire extension and made extinguishment difficult.

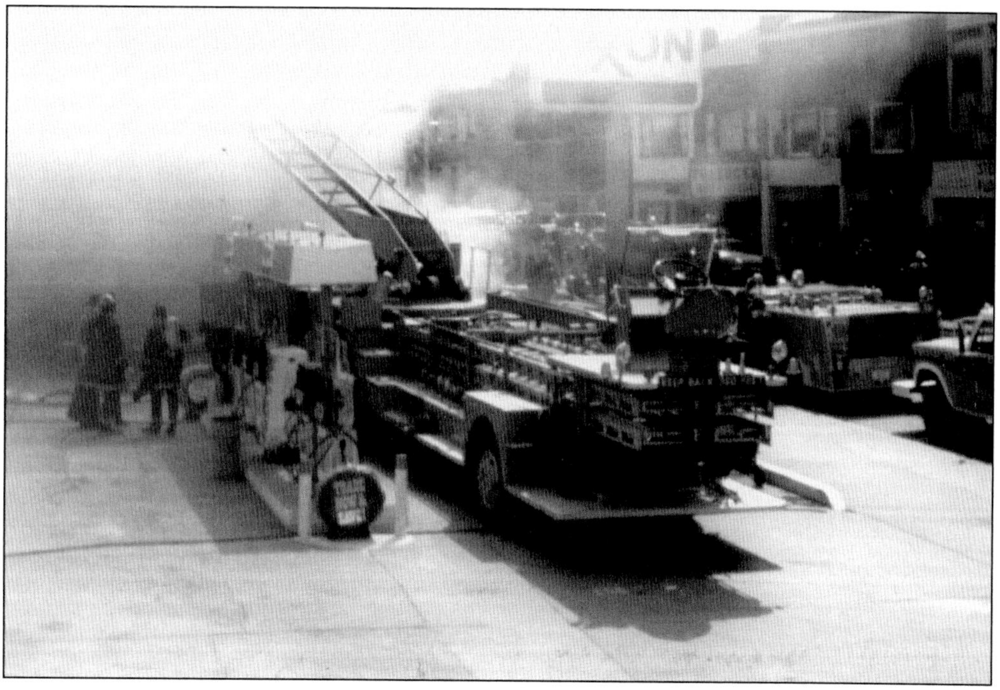

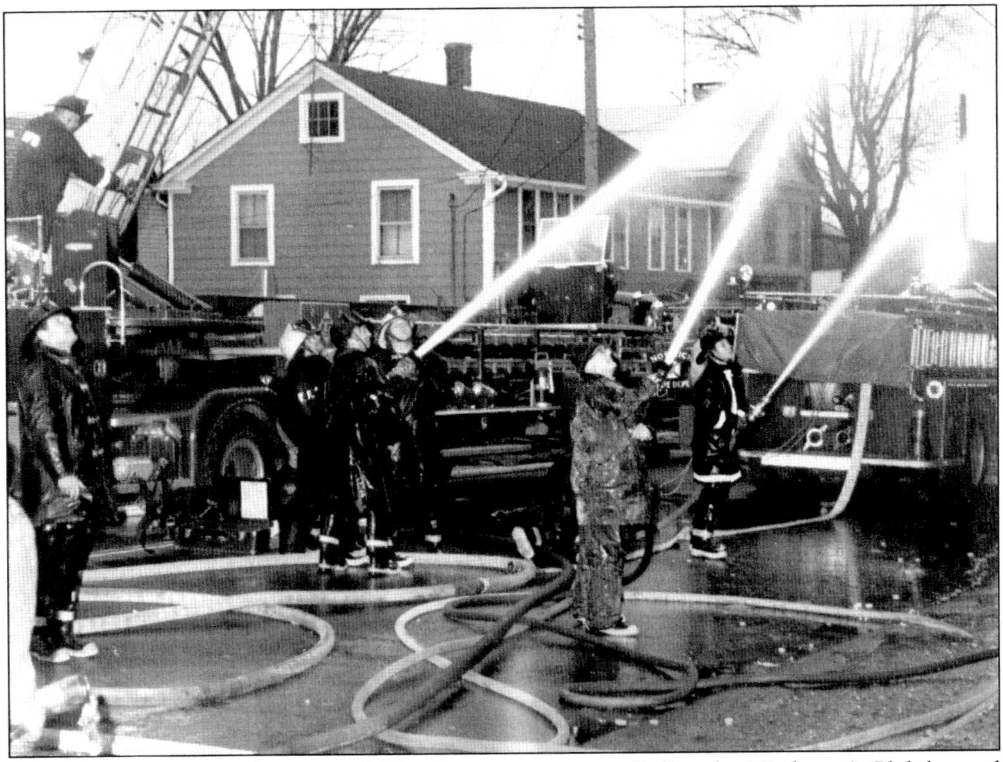

On December 1, 1974, the Norwich Fire Department was called to the Workmen's Club located at the corner of Sixth Street and North Main Street. The fire, which was started by a faulty television set, consumed most of the third floor and threatened two adjoining apartments.

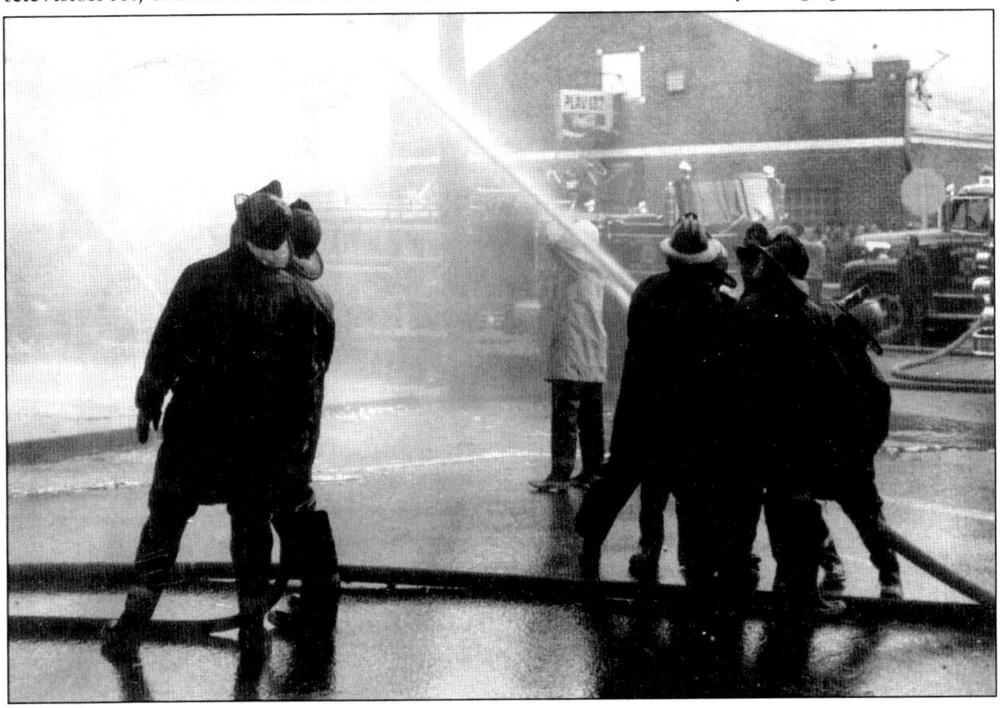

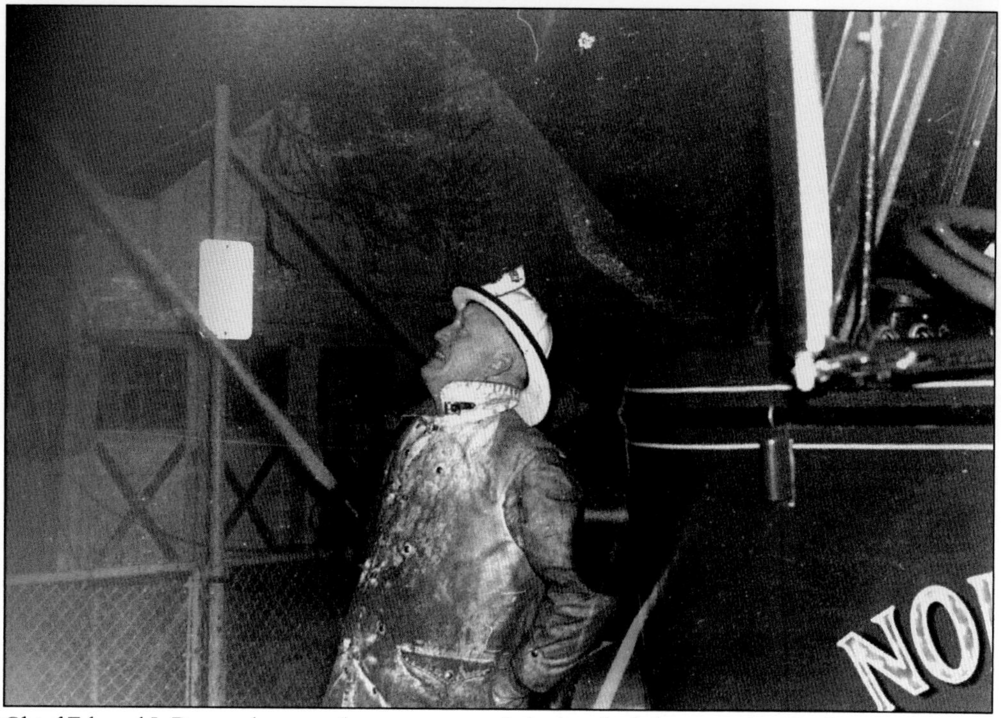

Chief Edward J. Breen observes the progress made by his firefighters at the Workmen's Club fire on North Main Street.

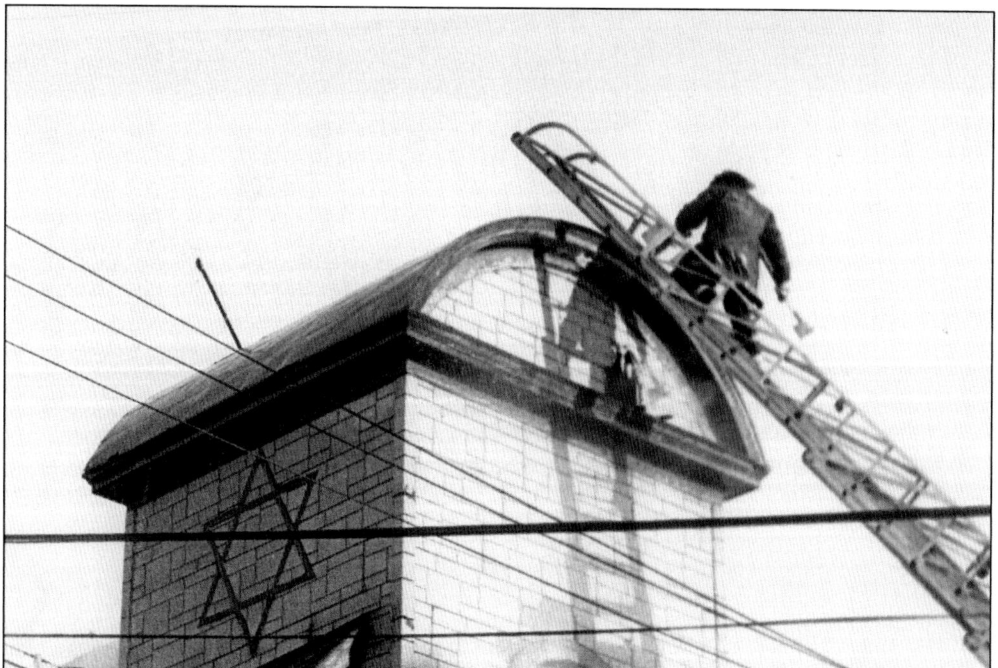

An unidentified fireman creates a ventilation hole in the former west side synagogue located at 40 High Street on December 31, 1974. The fire, which began in an adjacent three-story, wood-framed structure, spread to the former synagogue with the help of high winds.

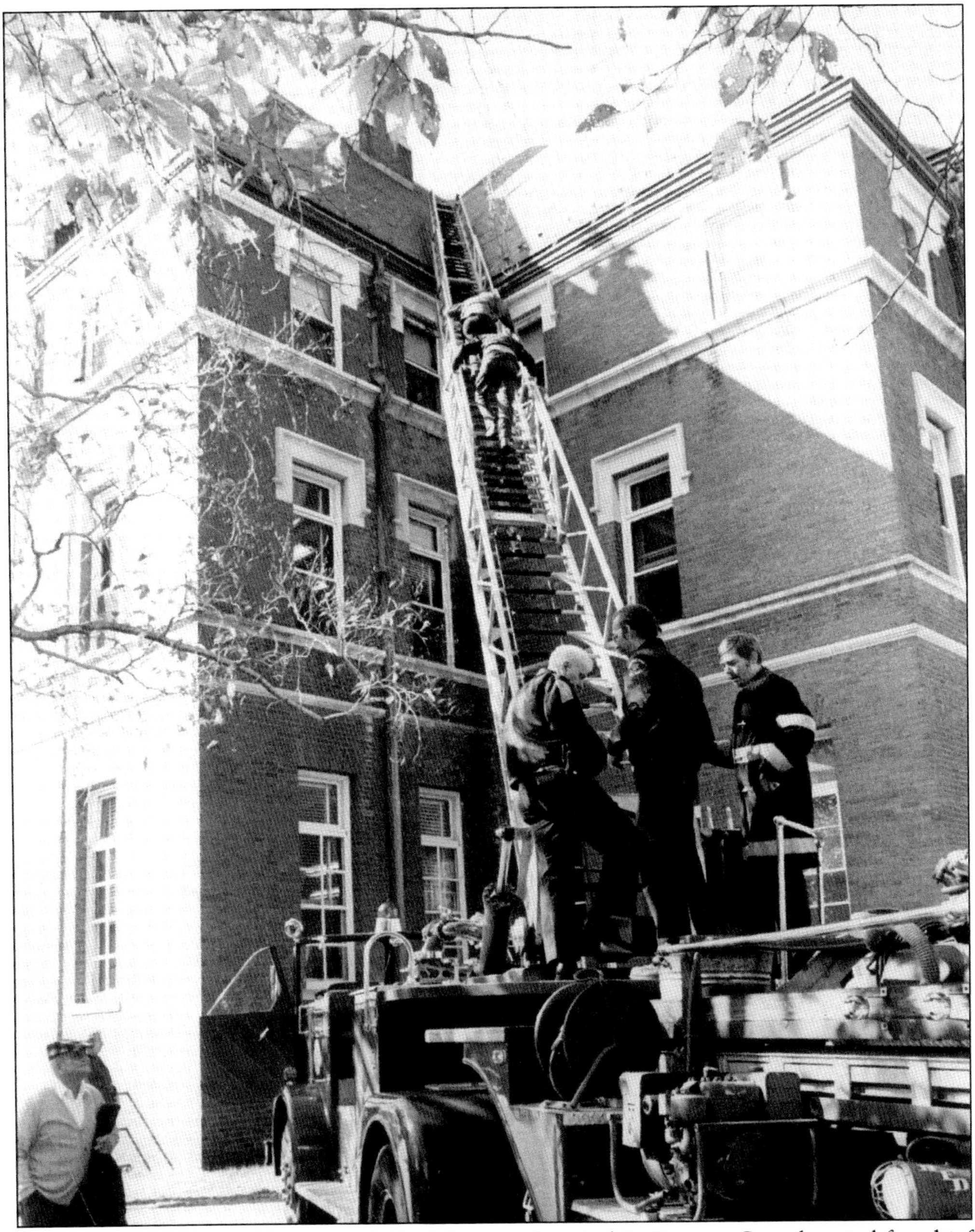

Ladder No. 1 is pictured here at a drill of the state hospital in Preston. State hospital fire chief William Dougherty (left) greets firefighters Stanley Homski Jr. (center) and William Dennis (right) on the turntable of ladder No. 1. Before taking the state hospital fire chief position, William Dougherty completed a career with the Norwich Fire Department.

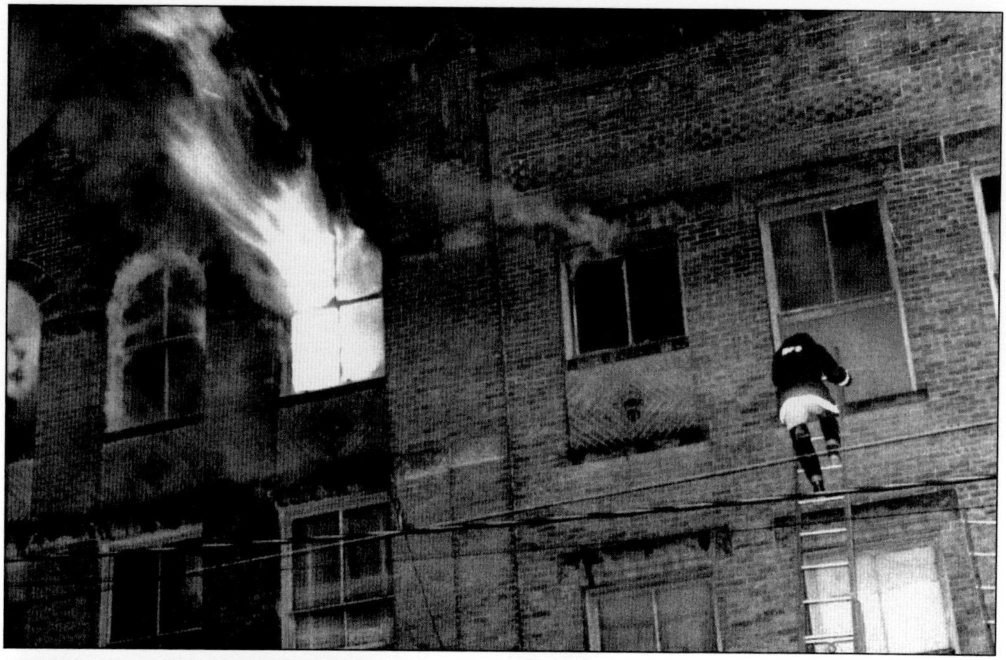

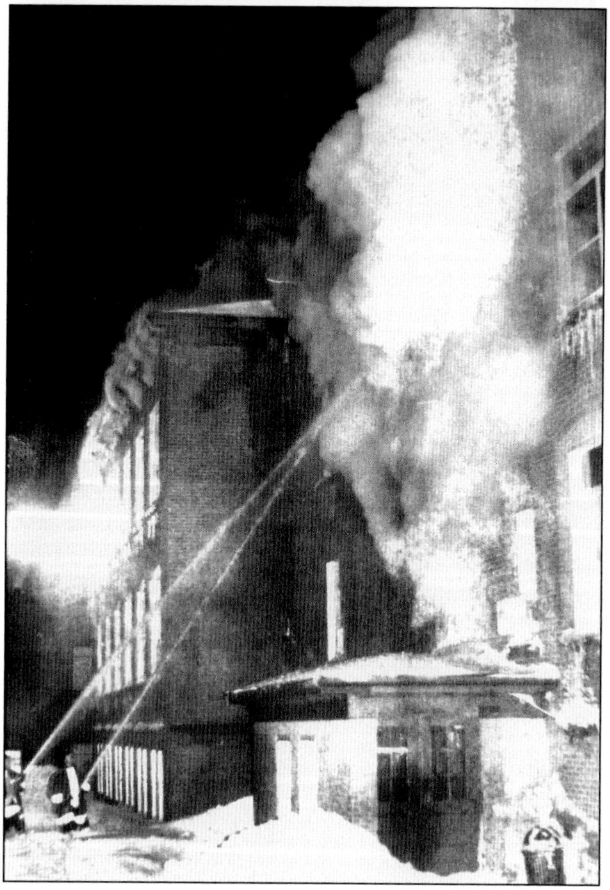

The Bradlaw building of the Norwich Free Academy received heavy damage following a fire that broke out during a 1978 snowstorm. Fire crews experienced impassable snowdrifts while responding to the alarm. Town plow trucks were needed to get the fire apparatus near the building.

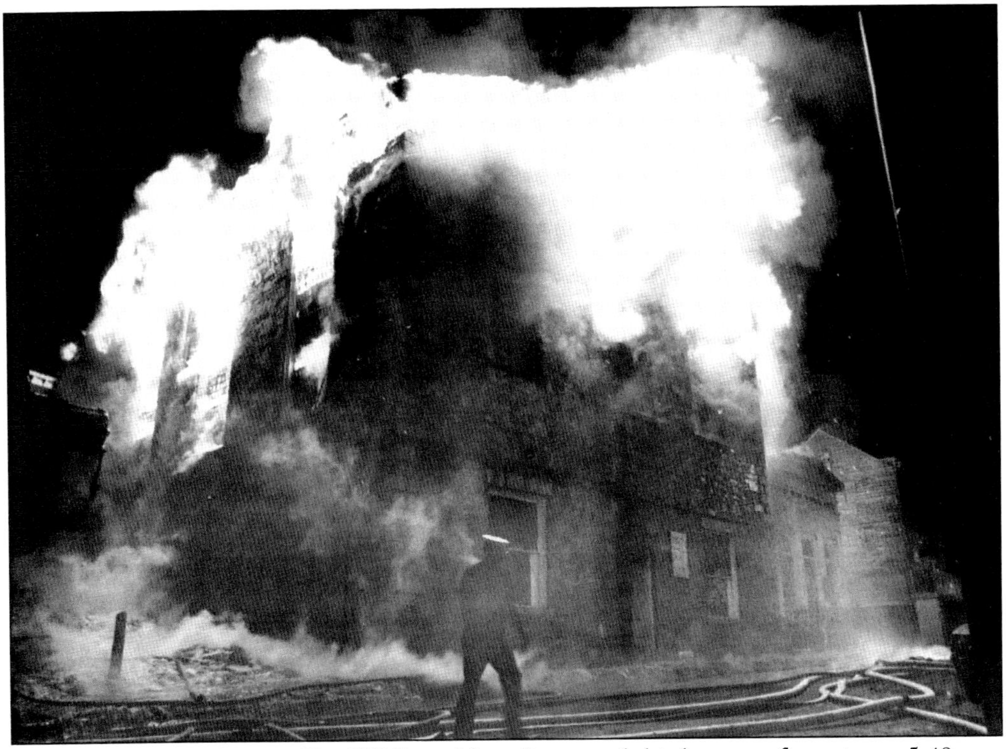

On March 7, 1978, Deputy Chief William Silvia discovered this large conflagration at 5:48 p.m. while driving through downtown. The former Episcopal church, located at 36 Main Street, quickly went up in flames as firemen arrived. A defensive attack was put into motion but not before a Norwich fireman was injured from a large piece of falling debris.

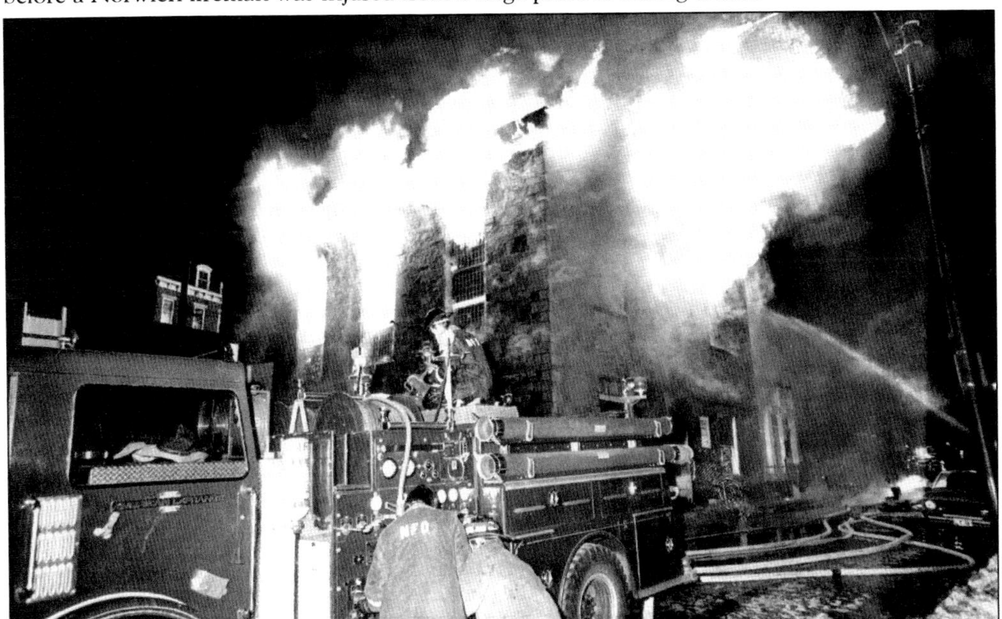

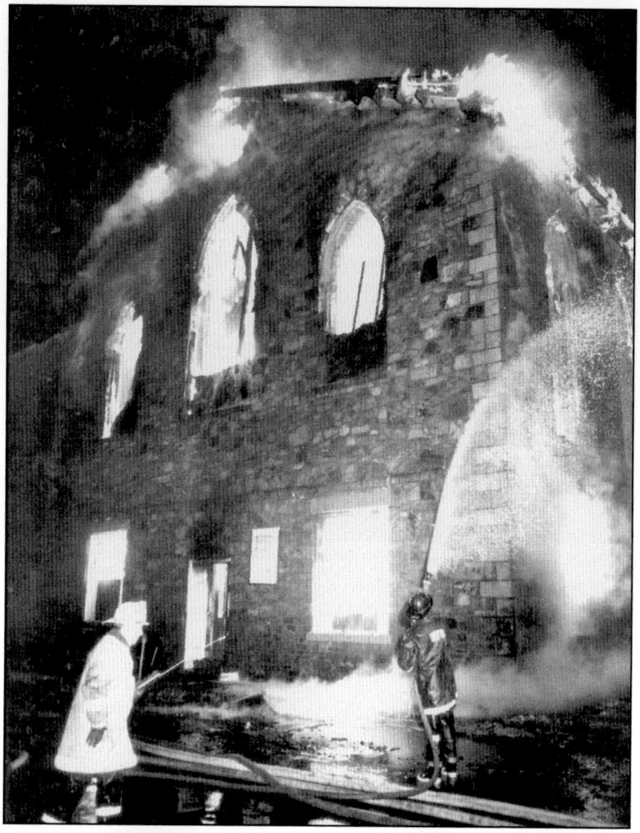

Chief Harold Lamphere Jr. (left) arrives on scene as defensive operations are put into place. Second and third alarms were eventually ordered to help contain the blaze. Below, the morning sunlight reveals the skeleton of the 150-year-old church.

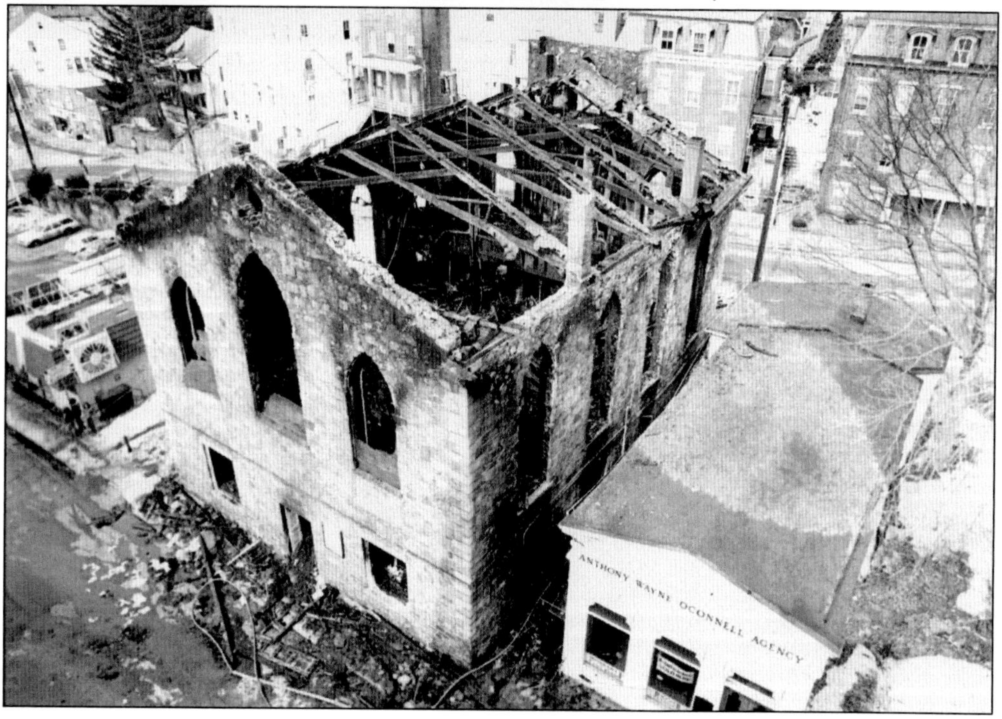

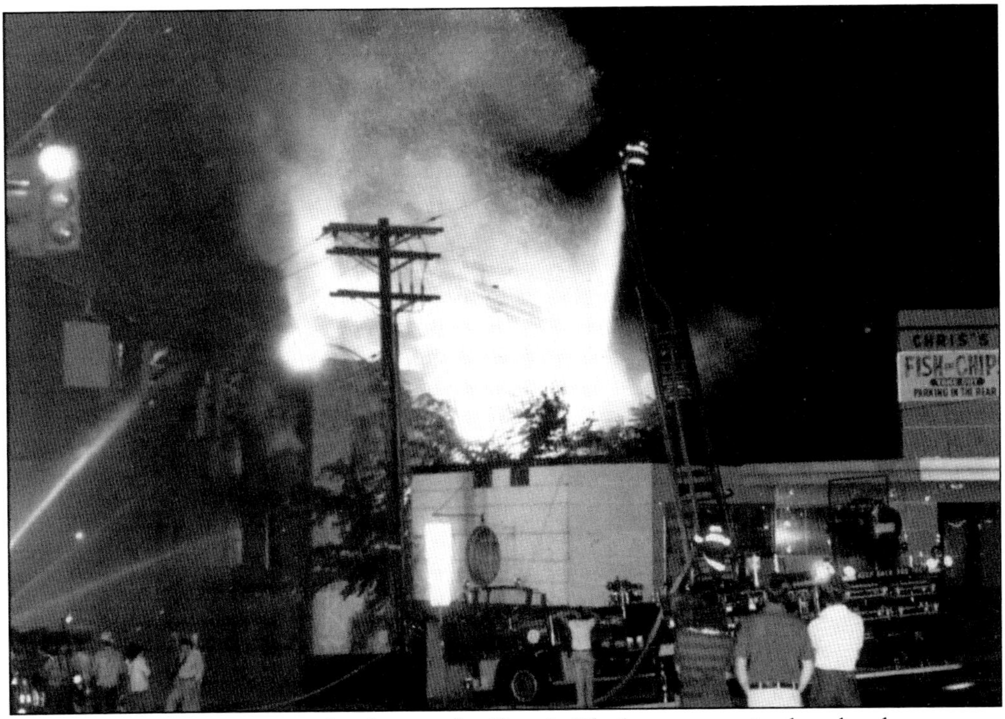

On June 23, 1979, reports of a fire in the Pasnik Block were received at headquarters at 11:00 p.m. The large three-story brick apartment building was located at 19 West Main Street near Washington Square. Firefighters battled the intense blaze with aerial master streams and large hand lines.

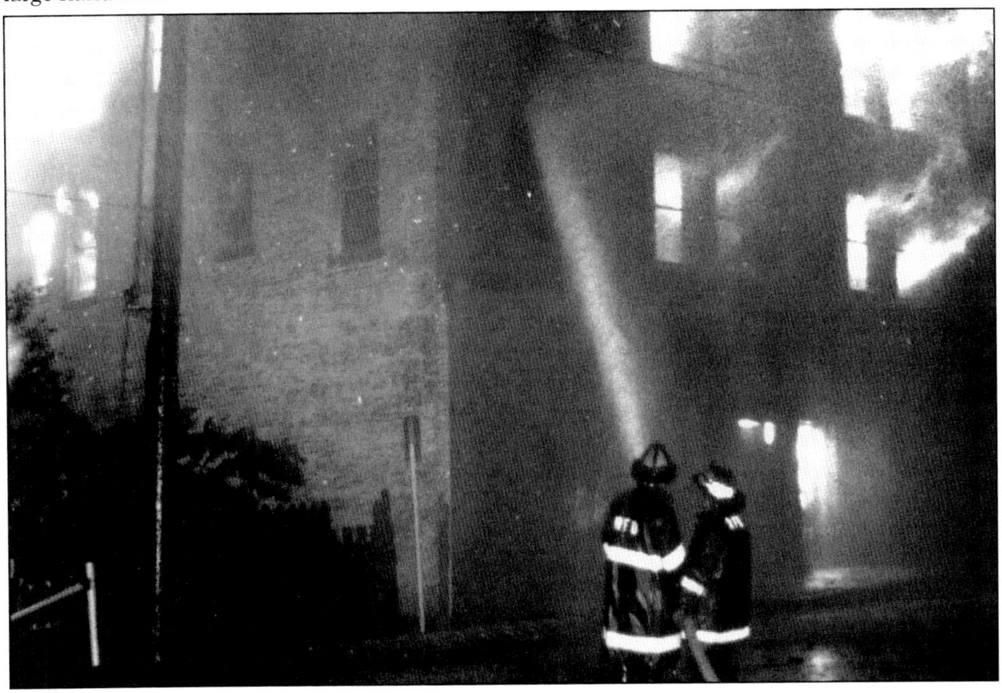

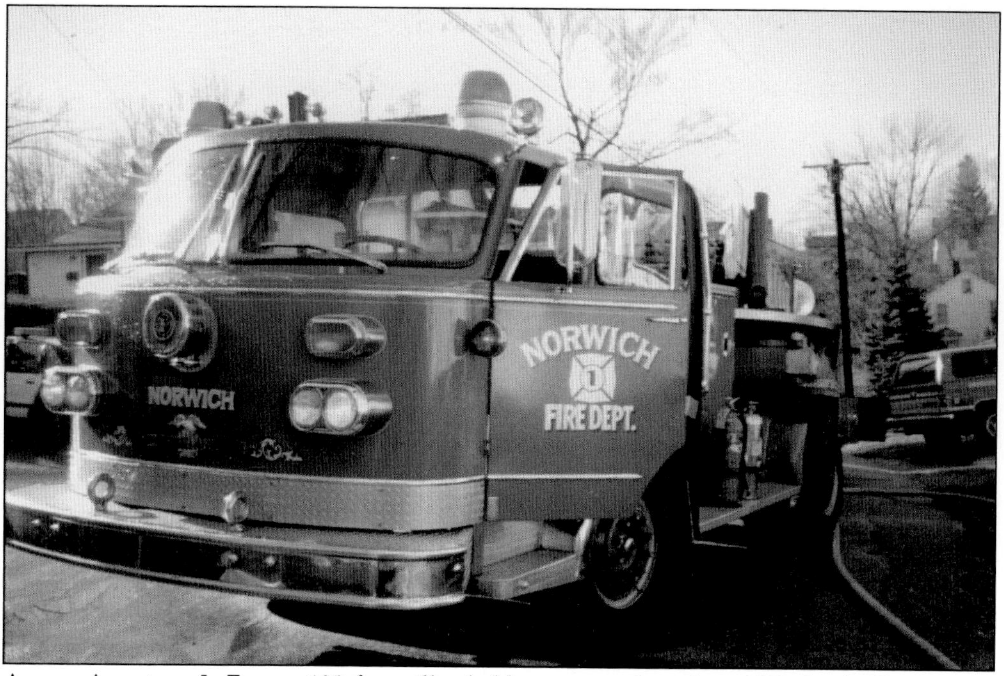

A new American LaFrance 100-foot tiller ladder was purchased in 1979 for $165,000. The 350-horsepower motor with automatic transmission was drastically different than the manual transmissions of the older apparatus. The new aerial was designated ladder No. 1 while the old ladder was sent to Greeneville's firehouse as a reserve. Below, firefighters are shown the proper setup for ladder pipe operations.

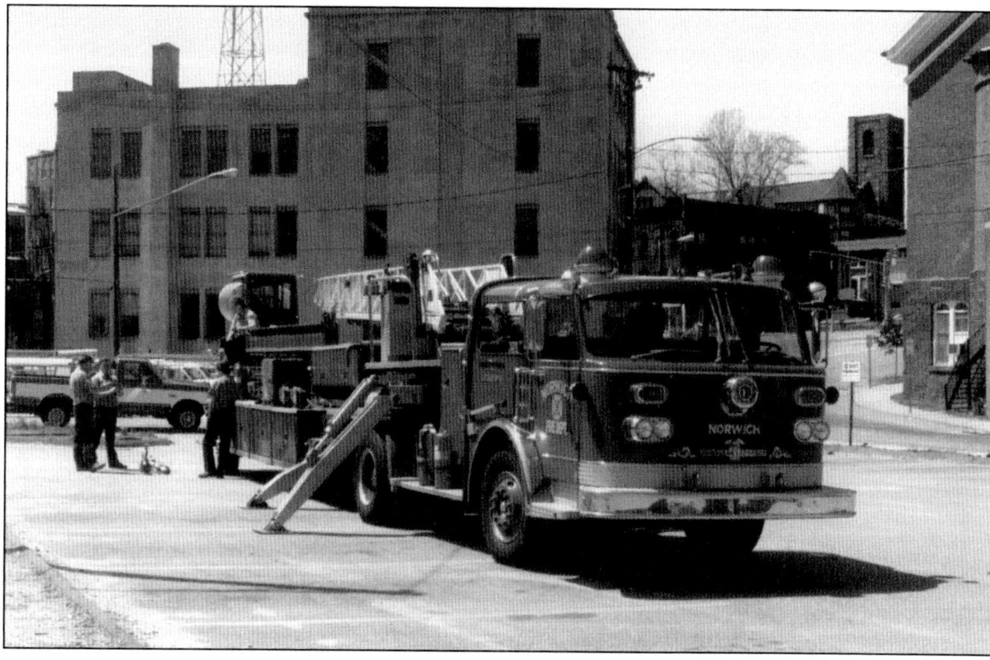

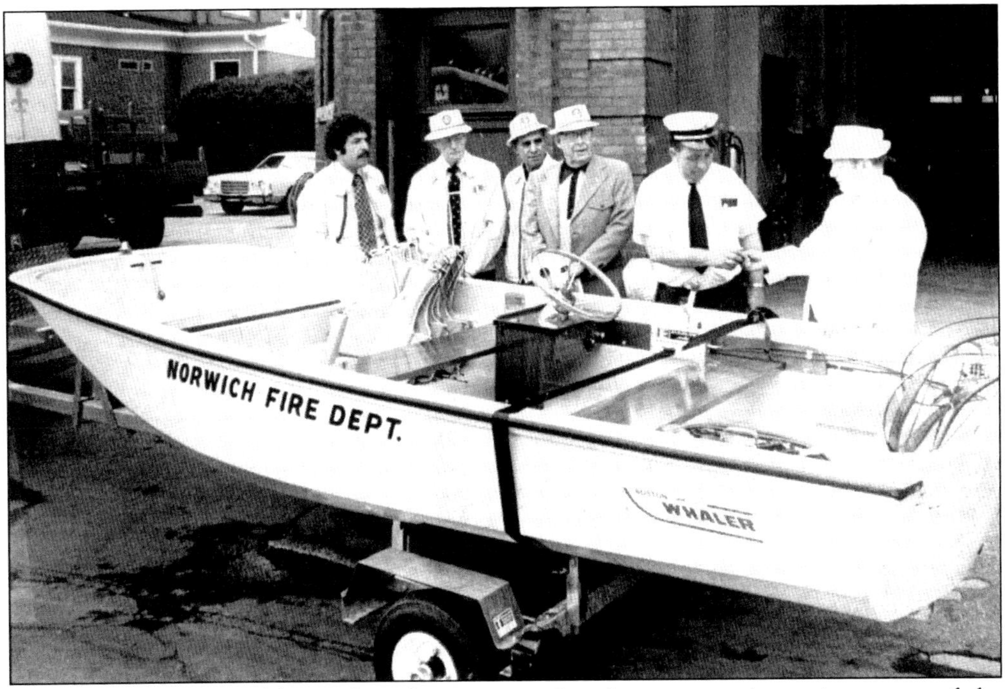

Above, Fire Chief Harold Lamphere Jr. is presented with a new marine unit courtesy of the Norwich Lions Club. The 14-foot Boston Whaler was designated marine No. 1. Below, Pvt. John Mathieu (left) and Pvt. John Langford (second from left) receive their Coast Guard safety check.

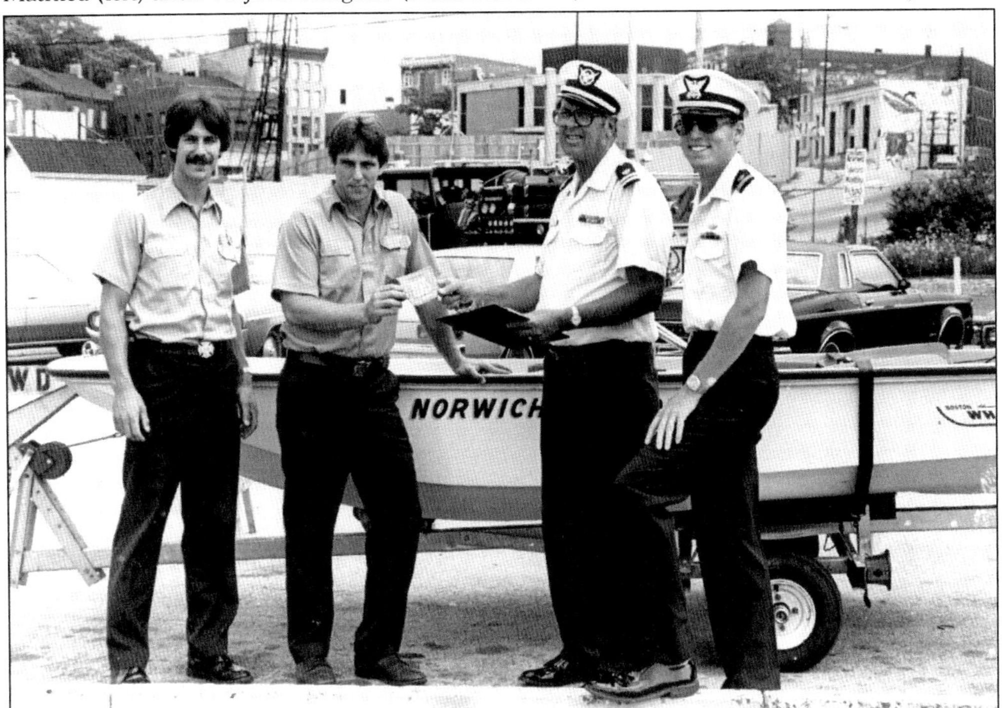

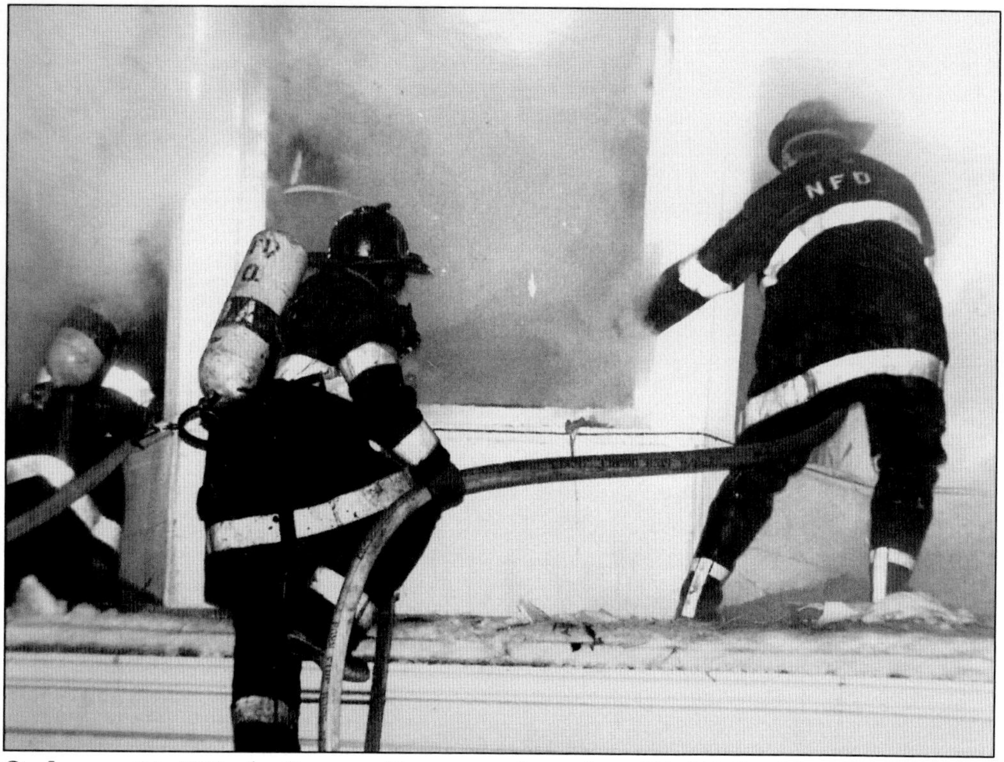

On January 24, 1982, the Stavrous Restaurant, located at 184–198 Franklin Street went up in flames. The three-story wooden structure was heavily damaged by fire. Below, firefighters establish a master stream operation to combat the flames.

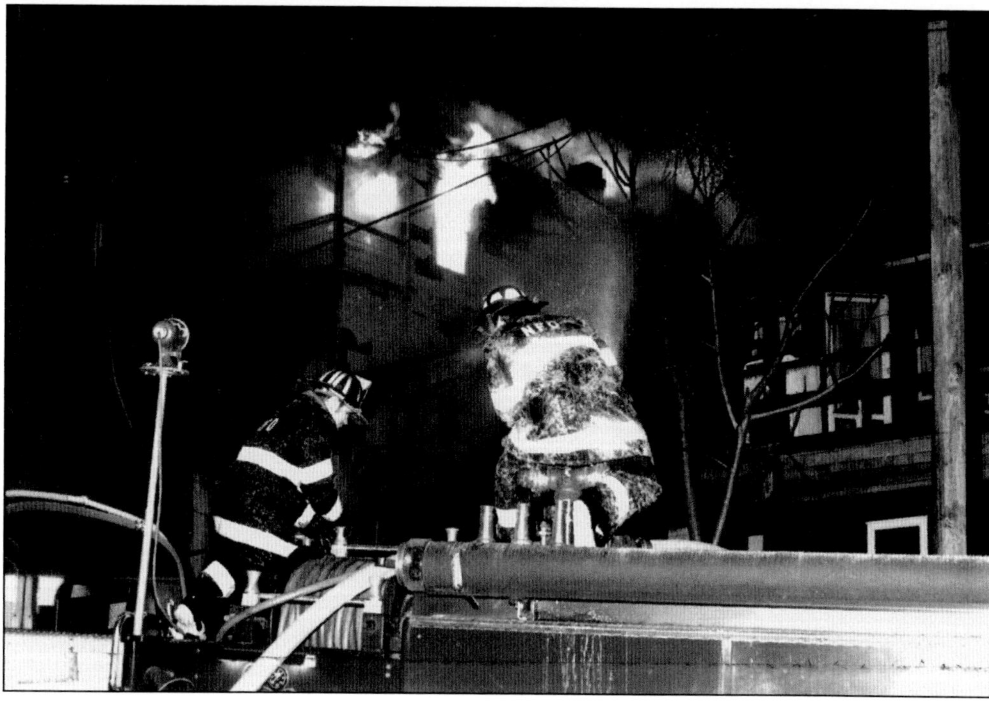

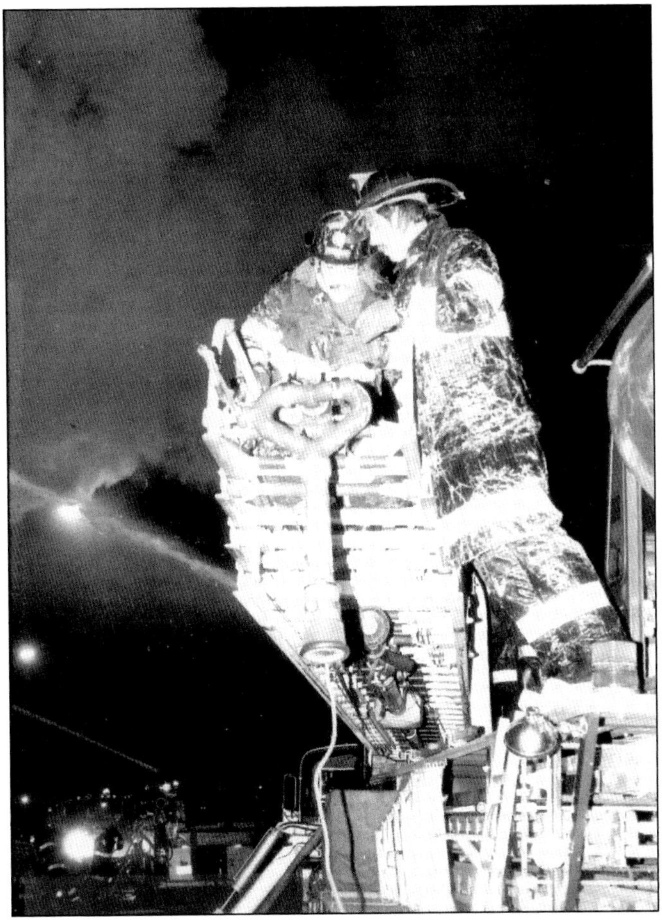

At right, firefighters ignore the bitter cold as they assemble the ladder pipe. Below, the subzero temperatures that were experienced during firefighting operations created thick layers of ice on the apparatus.

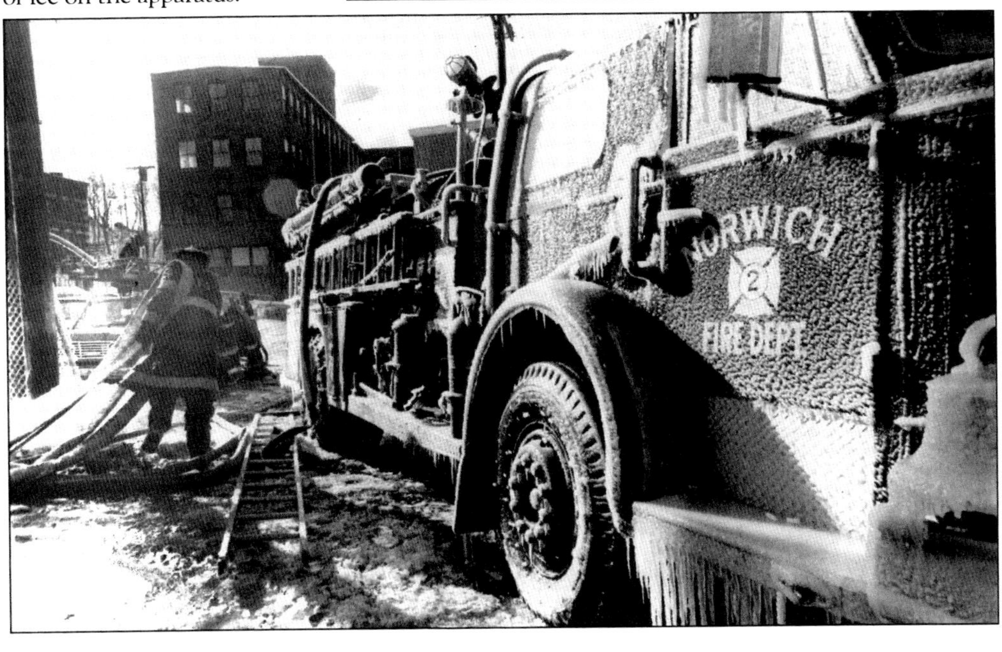

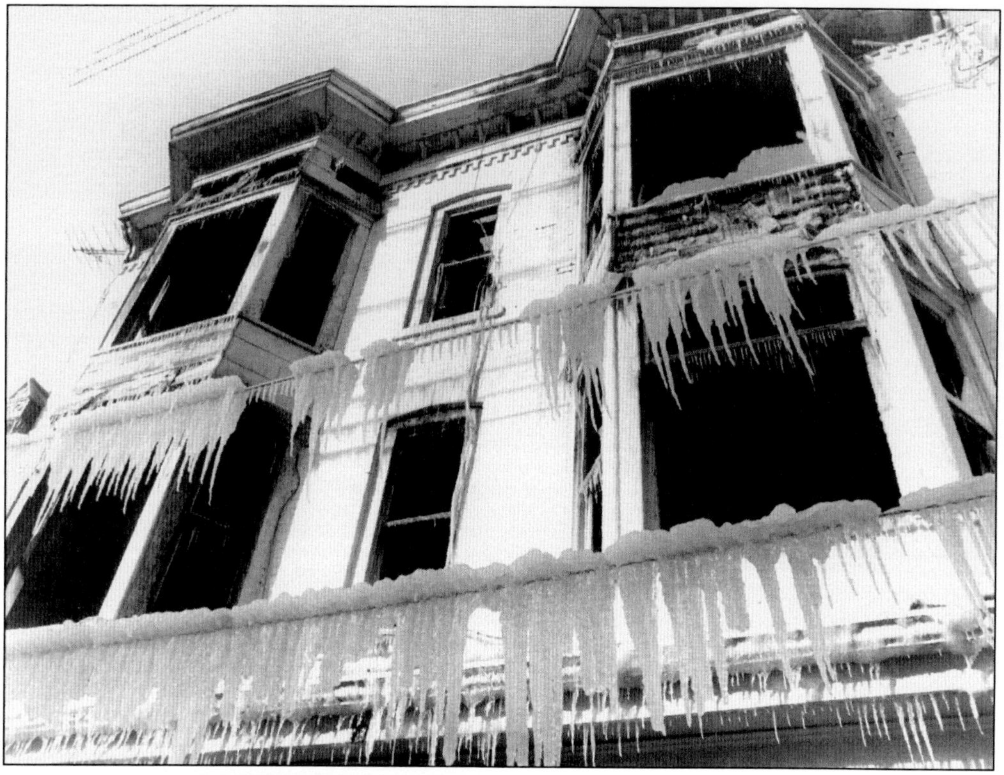

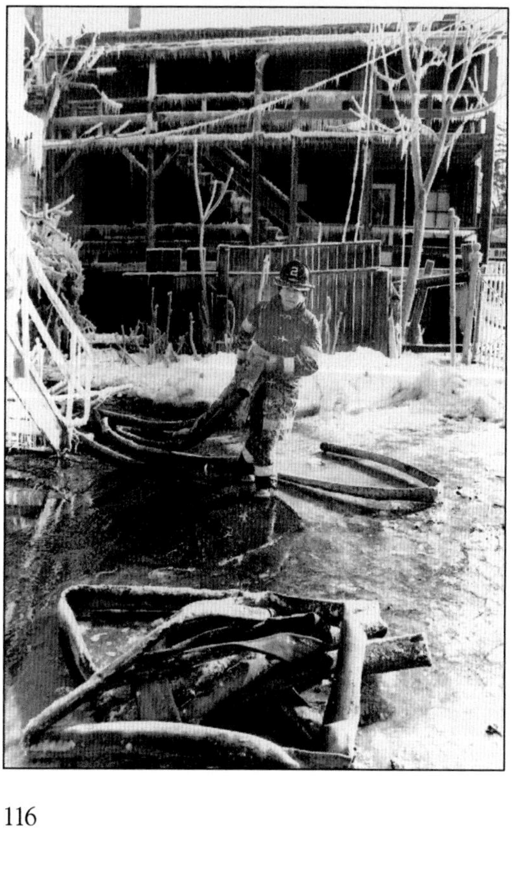

Above, approximately two dozen residents were displaced from their apartments following the Stavrous fire on Franklin Street. At left, firefighter Fred Haige is seen dragging partially frozen hose following the fire.

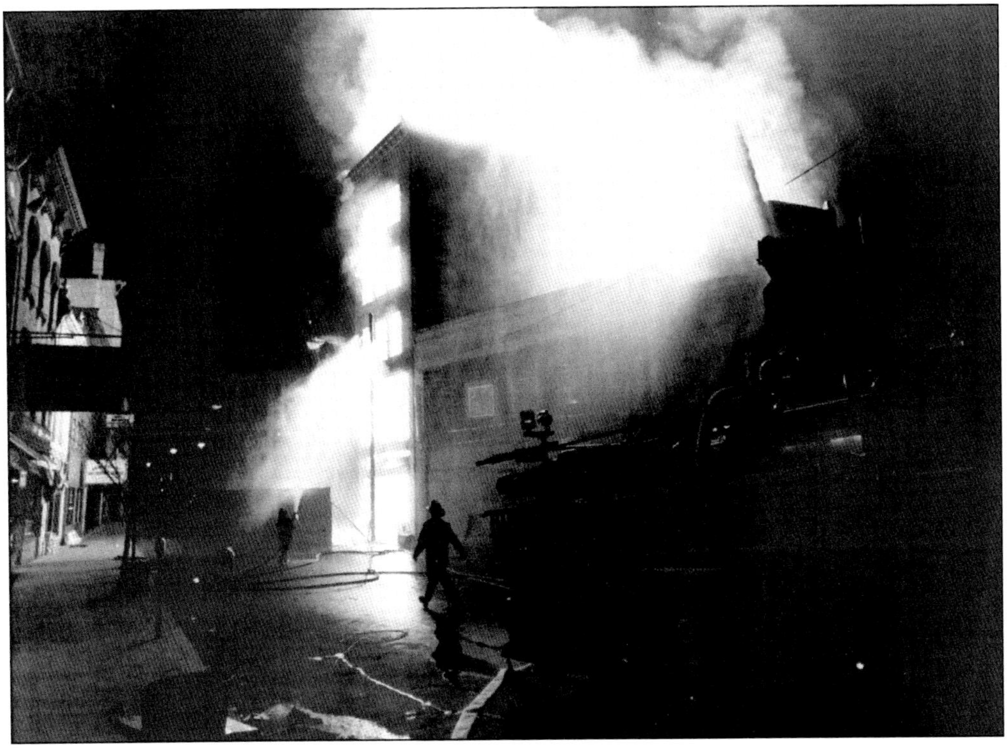

On April 22, 1982, a large fire consumed a five-story block in the downtown area. Slated for demolition, the building had already been cleared of debris and the utilities removed. Firefighters operated master streams and hand lines from Main Street.

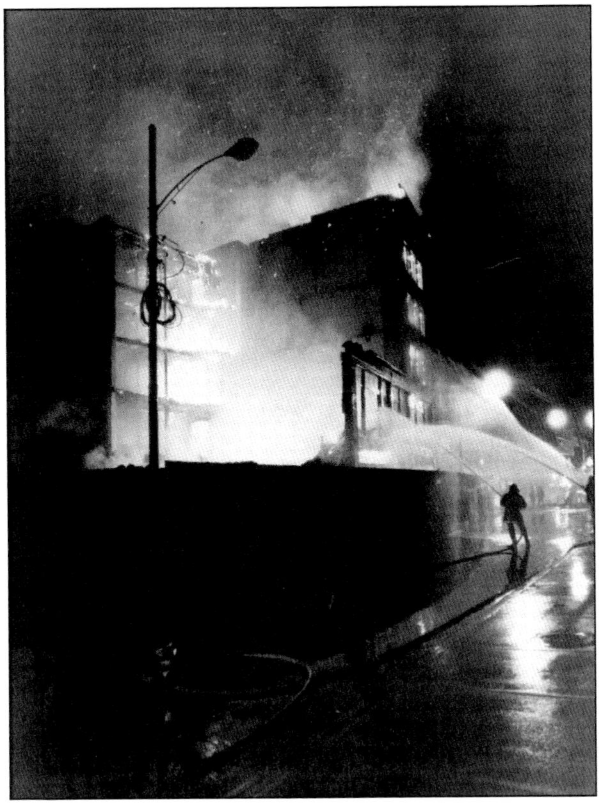

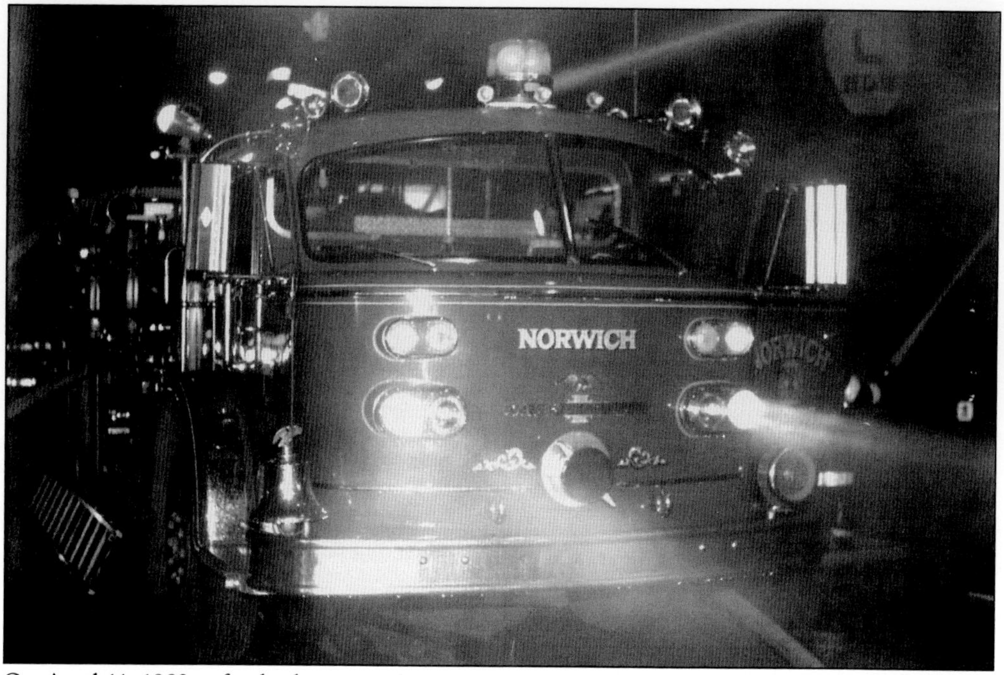

On April 11, 1982, a fire broke out at 6:21 p.m. in Leo's Café and quickly spread to the adjacent bowling alley and auto parts store on North Main Street. Above, firefighter William Dennis operates a master stream on the blaze. The aftermath of the fire was clearly visible the following day (below).

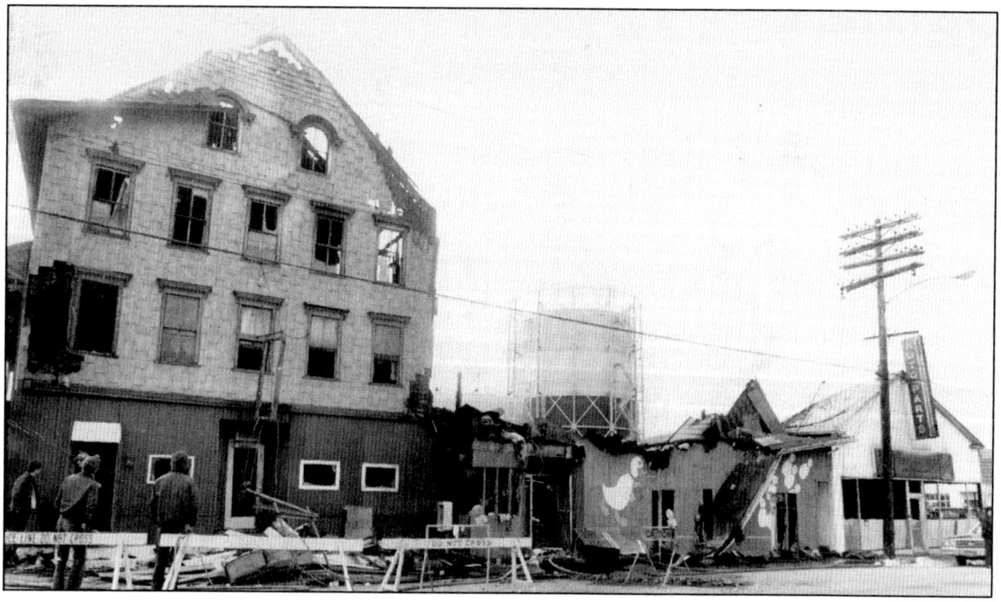

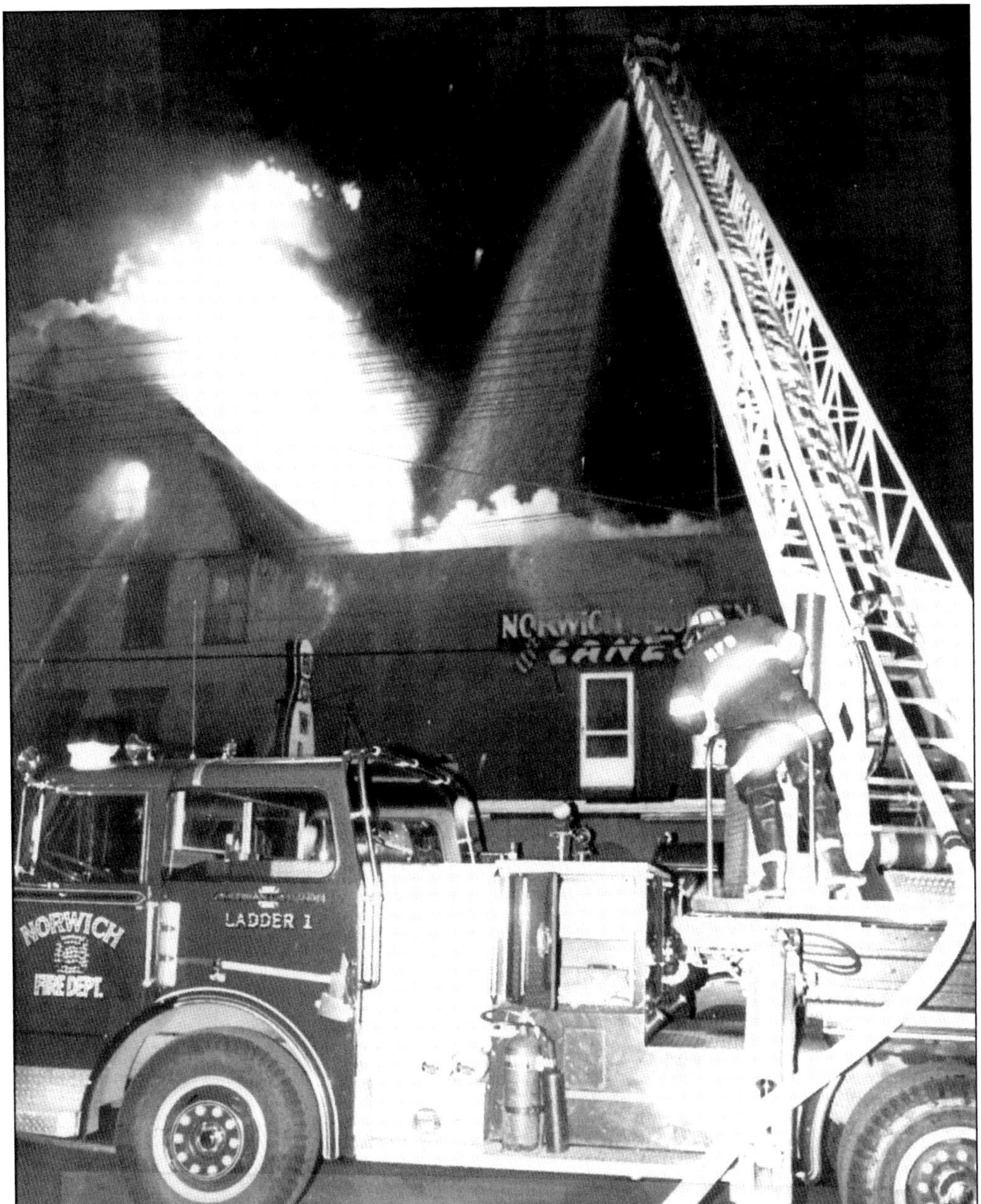

Ladder No. 1 is seen here operating an aerial master stream above the power lines of North Main Street. The Norwich Lanes fire proved to be difficult because of the close proximity of the power lines, the bowstring-truss roof construction of the bowling alley, and the mixed flammable and combustible contents of the auto body shop.

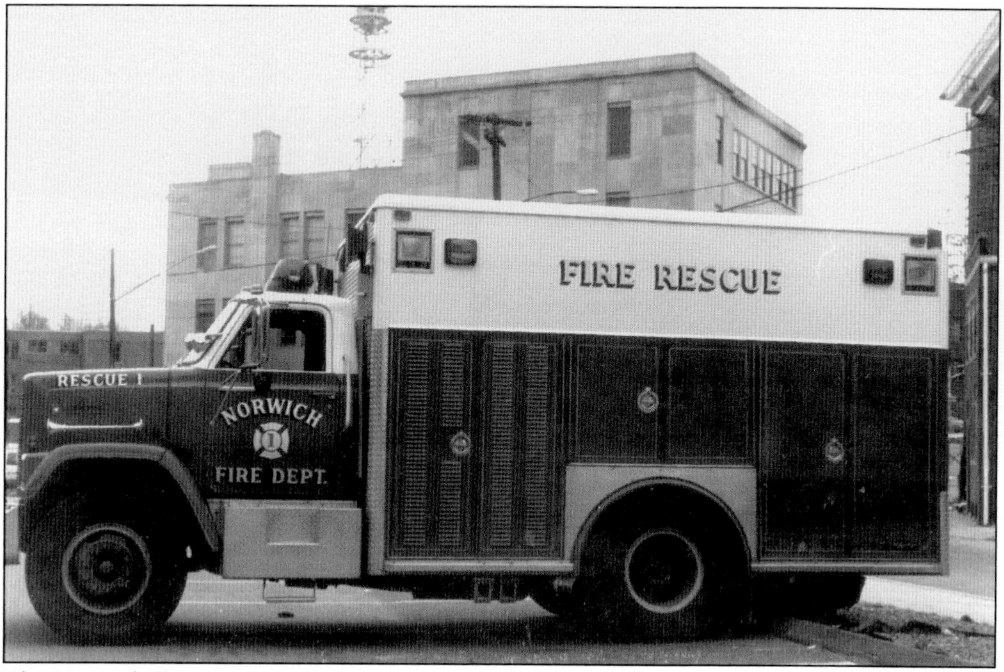

The Norwich Fire Department received a new heavy rescue in 1984. The GMC Ranger was operated as a two-man company until 1995 when the department was consolidated.

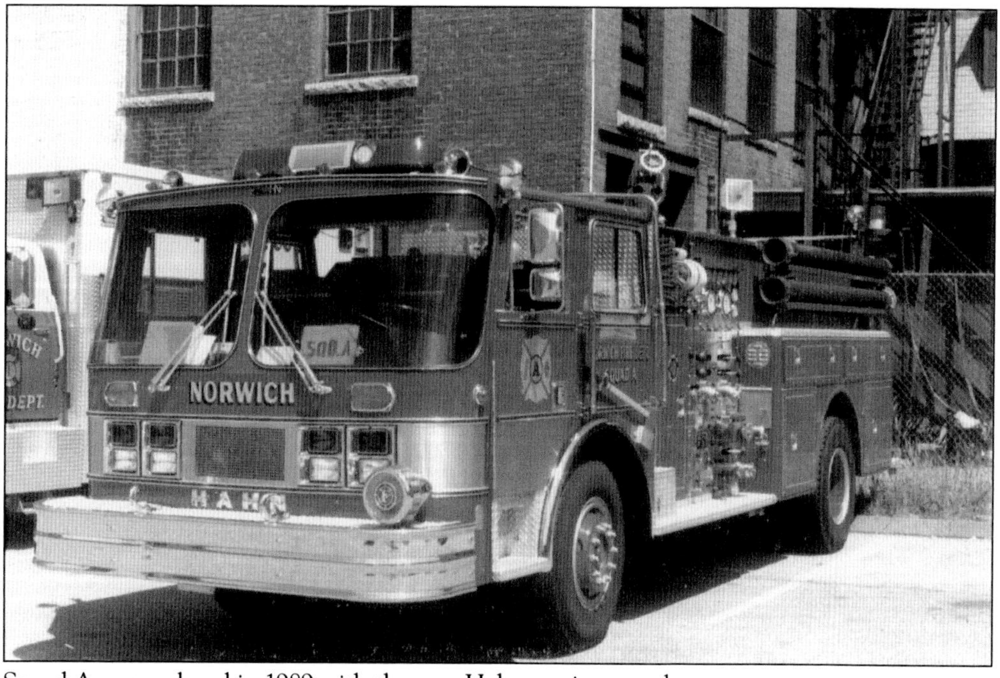

Squad A was replaced in 1989 with the new Hahn engine seen here.

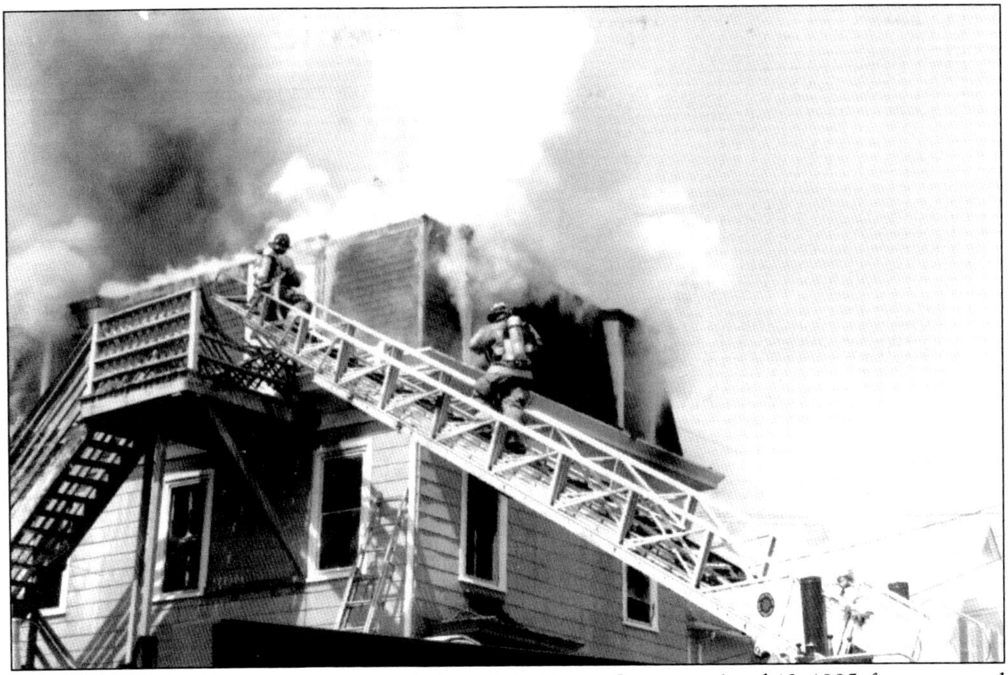

The Norwich Fire Department responded to 30–32 Grove Street on April 12, 1995, for a reported structure fire. Firemen work off ladder No. 1 while attempting to establish ventilation in this three-story mansard Victorian.

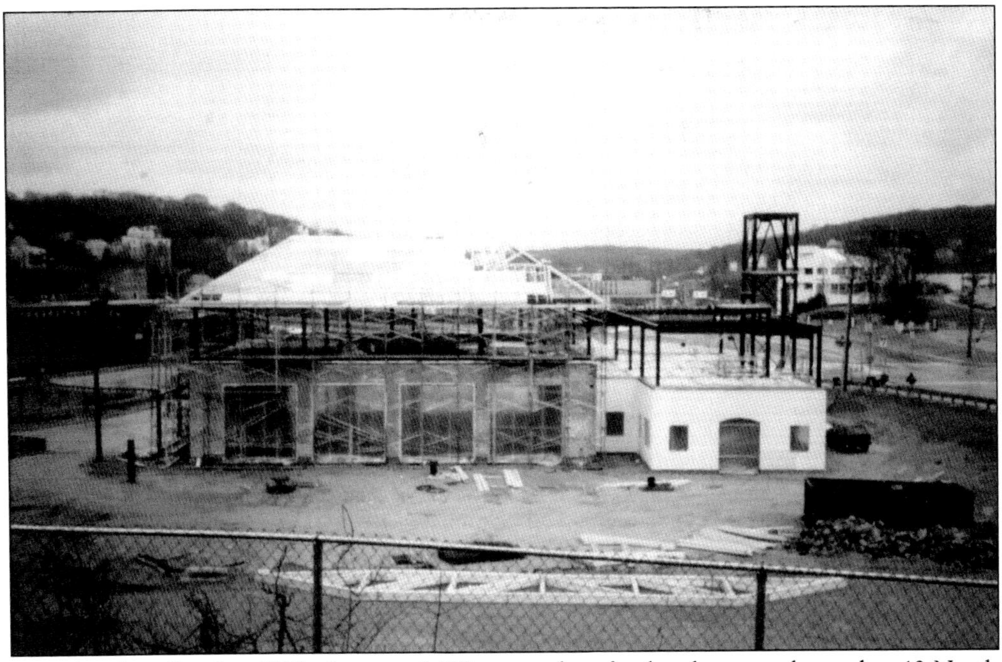

Completed in October 1995, the new 6,400-square-foot fire headquarters located at 10 North Thames Street was designed to house multiple fire companies.

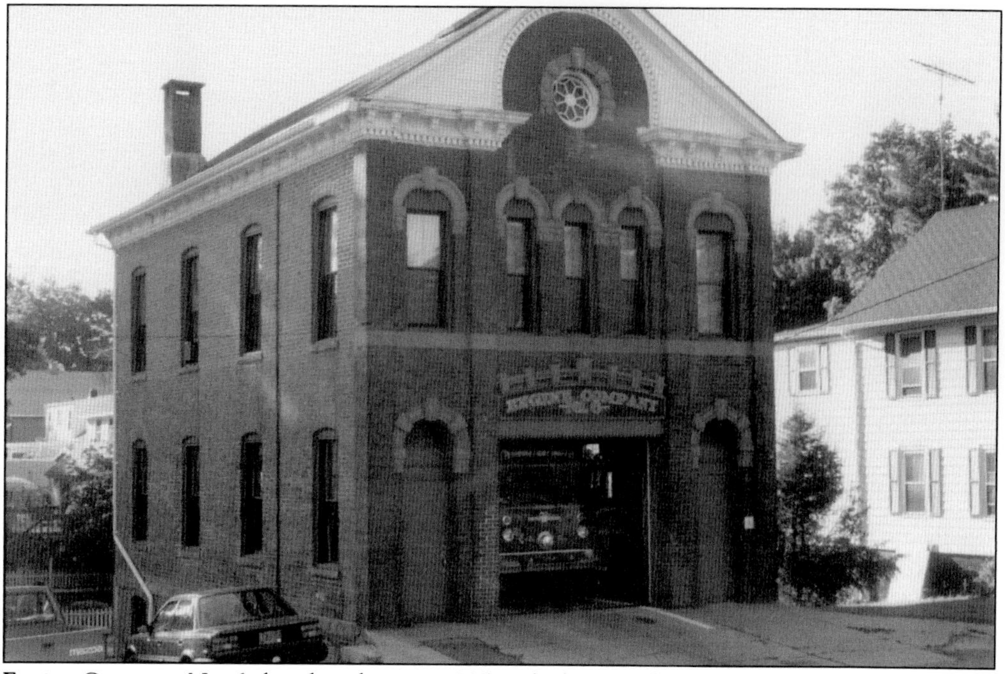

Engine Company No. 6 closed its doors in 1995 with the completion of the new headquarters at 10 North Thames Street.

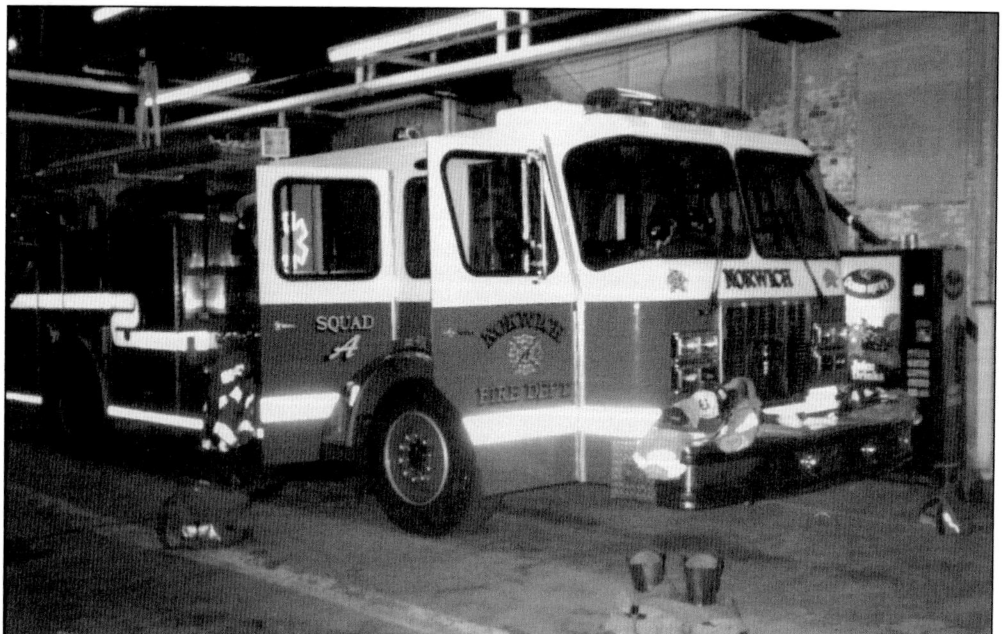

Squad A continued to operate from the Chestnut Street firehouse following the completion of the new headquarters in 1995. In 1998, the old headquarters on Chestnut Street was closed after 94 years of continuous service.

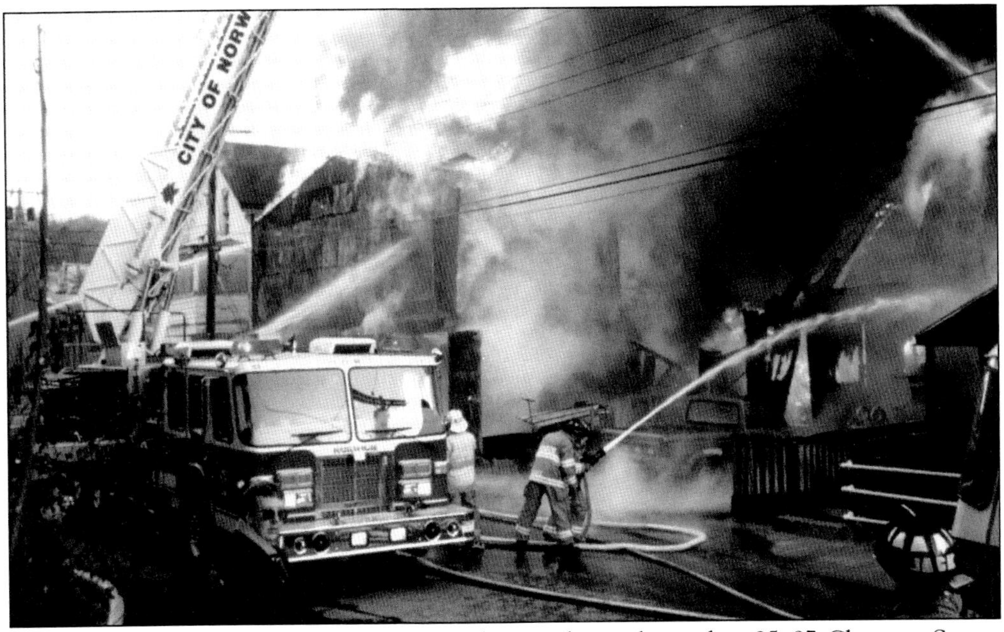

On January 3, 1999, a large blaze consumed a warehouse located at 95–97 Chestnut Street. The fire proved to be difficult because of a frozen hydrant, low water pressure, and numerous hazardous materials stored on-site.

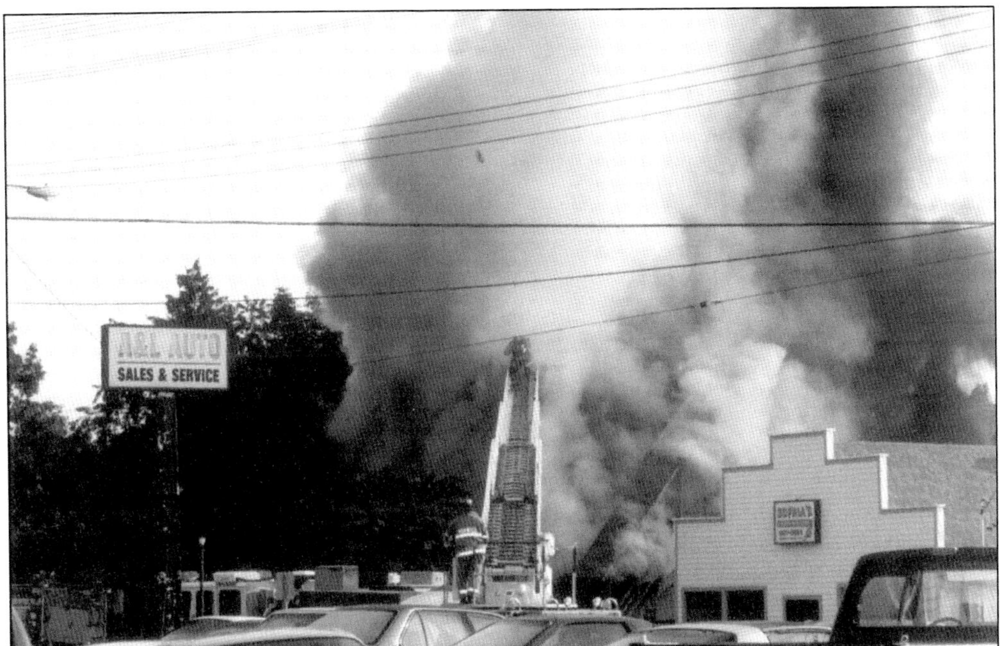

On October 4, 2000, units responded to 635 North Main Street for a reported structure fire. Upon arrival, department personnel discovered heavy smoke conditions emanating from the rear of the building. Truck No. 1's bucket is just visible in the dense smoke.

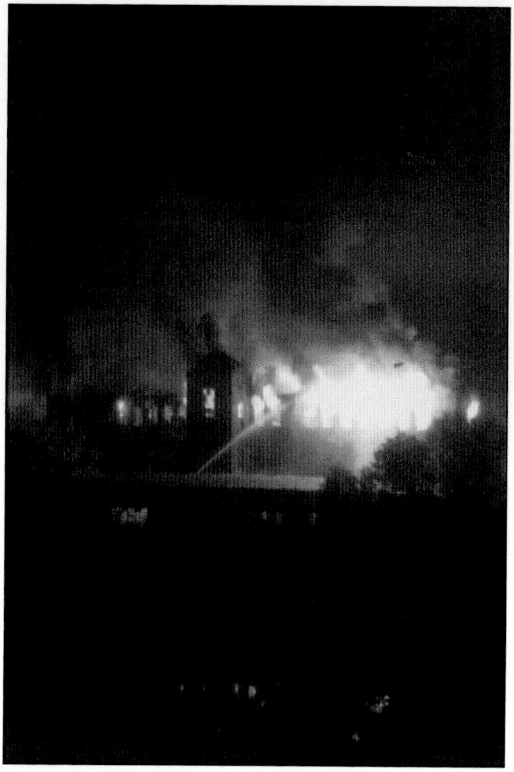

Norwich units responded to 18 Falls Avenue on September 13, 2002, for a reported structure fire. Upon leaving the station, responding units could see the former Glen Cairn mill on fire. This multialarm fire was caused by a juvenile with a flare gun.

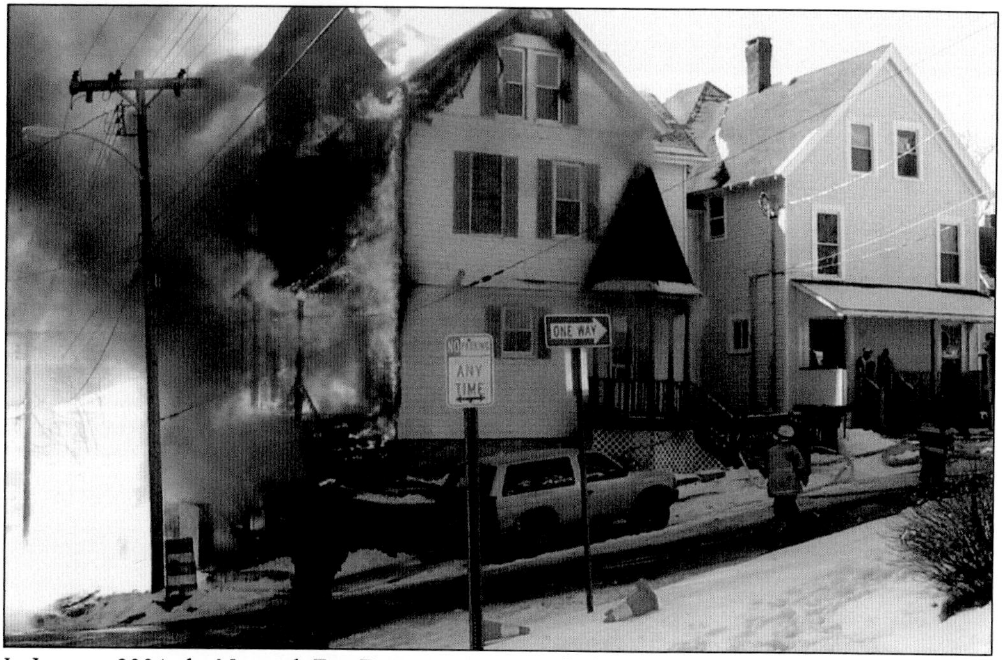

In January 2004, the Norwich Fire Department was called to the corner of Cliff and Park Streets for a structure fire. This two-and-a-half-story wood frame Victorian had made considerable headway prior to the arrival of the fire department.

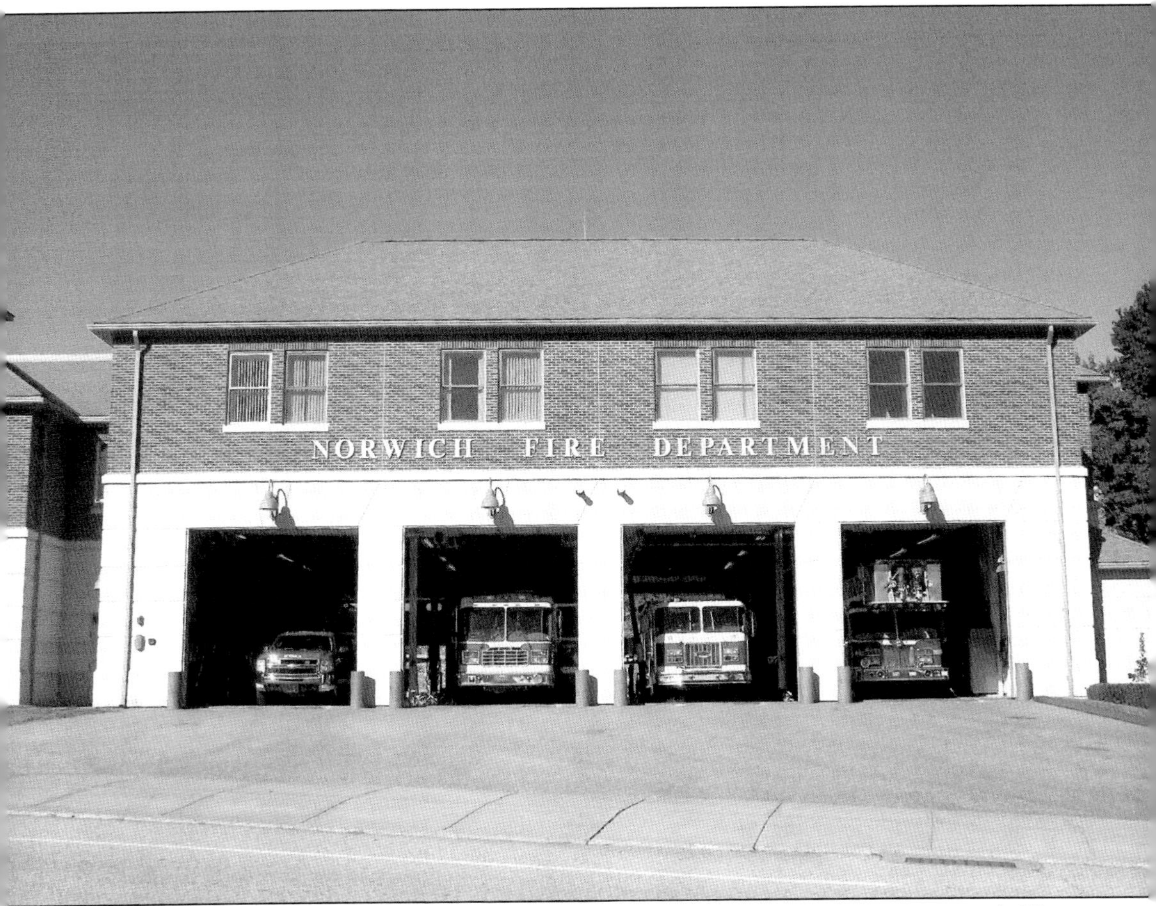

In 2007, the Norwich Fire Department added the battalion chief rank to its personnel. The department's staffing today consists of 36 privates, 8 lieutenants, 5 captains, 4 battalion chiefs, 2 fire inspectors, 1 safety/training marshal, 3 secretaries, and the chief. Seen above is headquarters with units, from left to right, battalion No. 1, squad A, engine No. 3, and truck No. 1.

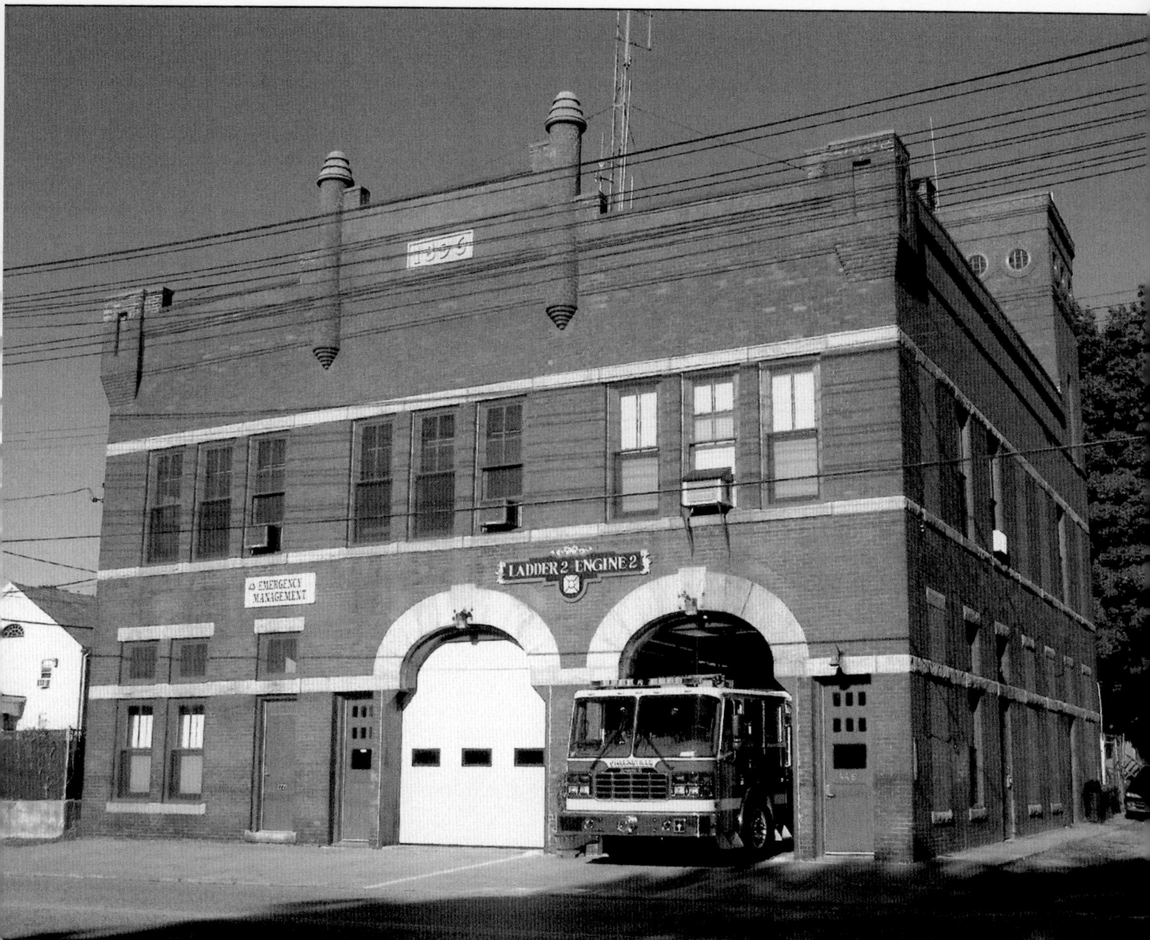

Retaining much of its original glory, Engine Company No. 2 continues to protect the Greeneville District of Norwich. In 2004–2005, the station received numerous updates. In 2007, a new Ferrera Inferno engine was purchase and placed in Greeneville. Station No. 2 is one of the oldest firehouses still in operation in the state of Connecticut.

Chiefs of the Norwich and Greeneville Fire Departments

Col. James Spaulding 1846*
Edward S. Osgood 1857*
Henry Peale 1860-1862
Sylvanus Downer 1862-1863
John Wilbur *(Greeneville)* 1862-1866
P. St. M. Andrews 1863-1866
William M.Williams 1866-1870
Paul B. Green *(Greeneville)* 1866-1868
James Conklin *(Greeneville)* 1868-1870
Joseph B. Carrier1870-1871
D.R. Holmes *(Greeneville)* 1870-1872
Daniel A. Delanoy 1871-1873
George Johnson *(Greeneville)* 1872-1875
Joseph B. Carrier1873-1890

Consolidation of the Departments in 1876
Robert A. France *(Greeneville)* 1875-1876
Louis W. Greenberg 1890-1901
Howard L. Stanton 1901 -1929
Henry R. Taft 1929-1939
Joseph P. Tracey 1939-1952
Charles F. Matthews 1952-1953
William B. Confrey 1953-1967
Charles F. Matthews 1967-1967
Henry Milewski 1968-1971
Edward J. Breen 1971-1977
Harold H. Lamphere Jr. 1977-1989
Thomas W. Kirby 1989-1996
James F. Walsh 1996-2004
Kenneth Scandariato 2005-Present

*** Due to limited publications of the Norwich City Directory, the date of appointment and date of resignation are not known.**

The chiefs of the Norwich and Greeneville Fire Departments are listed here.

ACROSS AMERICA, PEOPLE ARE DISCOVERING SOMETHING WONDERFUL. *THEIR HERITAGE.*